PRACTICAL GUIDE ON SEASONAL ADJUSTMENT USING JDEMETRA+

DECEMBER 2025

ASIAN DEVELOPMENT BANK

ADB

6 ADB Avenue, Mandaluyong City, 1550 Metro Manila, Philippines
Tel +63 2 8632 4444; Fax +63 2 8636 2444
www.adb.org

ISBN 978-92-9277-611-4 (print); 978-92-9277-612-1 (PDF); 978-92-9277-613-8 (ebook)
Publication Stock No. TIM250534-2
DOI: http://dx.doi.org/10.22617/TIM250534-2

Cover design by Patricio Francisco.

CONTENTS

TABLES AND FIGURES

TABLES

FIGURES

ACKNOWLEDGMENTS

The Practical Guide to Seasonal Adjustment with JDemetra+ was prepared by the Data Division within the Economic Research and Development Impact Department at the Asian Development Bank. The Data Division wishes to recognize the contributions of the project team and also the statistical partners who made this publication a success. The Japan Fund for Prosperous and Resilient Asia and the Pacific provided financial support enabling the development of seasonal adjustment methodologies documented in the user guide.

We acknowledge the Statistical Committee of the Republic of Armenia and Kazakhstan Bureau of National Statistics for their active engagement and technical collaboration in advancing seasonal adjustment activities. Their openness to innovation and commitment to improving the quality and timeliness of official statistics have been vital in ensuring that economic indicators more accurately reflect turning points within the countries.

We sincerely appreciate the unwavering support from Statistical Committee of the Republic of Armenia particularly Stepan Mnatsakanyan (president); Anahit Safyan and Gagik Ananyan (deputy presidents); Lilit Petrosyan (board member); Lusya Khachatryan, Seda Movsisyan, Arus Tonapetyan, Susanna Khachatryan, Anna Vardanyan, Kristina Israelyan, Nane Dnoyan, and Naira Baboyan from the Macroeconomic Indicators and National Accounts Division; and Mariam Yeritsyan, Gayane Petrosyan, and Arpine Mkhitaryan from the Price Statistics and International Comparisons Division. Further, a big thank you to the Kazakhstan Bureau of National Statistics especially Turlubayev Maxat Kairatovich (head); Asset Nakipbekov, Zaure Zhunussova, and Gabiden Torebekov from the Department of National Accounts; Natalia Belonosova, Lyazzat Sailaukyzy, and Gulnar Bekisheva from the Department of Labor and Standard of Living Statistics; Aida Isabekova, Irina Torgautova, Zamzagul Dautova, and Alina Beisebekova from the Department of Production and Environmental Statistics; and Marina Misyura, Gulzhikhan Aigozina, Svetlana Grigorieva, Gulnur Begalina, and Saltanat Kairbekova from the Department of Services and Energy Statistics. Their participation played a crucial role in supporting evidence-based decision-making and policy formulation in both countries. For the pilot countries, seasonal adjustment removes predictable seasonal patterns and calendar effects, supporting more robust economic analysis, clarifying key turning points, and highlighting underlying trend dynamics. The seasonal adjustment approach is conducted in compliance with international standards, e.g., the International Monetary Fund Quarterly National Accounts Manual (2017) and the European Statistical System Guidelines on Seasonal Adjustment (2024). Continuous improvement follows from transparent documentation, quality diagnostics, and structured revision policies. Methodological soundness is maintained through a reference manual, comprehensive metadata, and regular quality assurance reviews.

The successful implementation of seasonal adjustment methodologies in Armenia and Kazakhstan was spearheaded by Stefan Schipper. The project team members composed of Pamela Lapitan, Gevorg Minasyan, Aizhan Bolatbayeva, and Alisher Tolepbergen provided support during the workshops, technical exchanges, and capacity-building efforts in Armenia and Kazakhstan. Brad Earvin Zuniga and Ma. Roselia Babalo provided administrative support throughout the implementation of the project.

The Asian Development Bank's partnership with Armenia and Kazakhstan is strengthening resilient, responsive statistical systems and promotes data-driven development across Central and West Asia and beyond. Together, we are building capacity, including in seasonal adjustment, that align with international standards to support consistent, transparent, and comparable statistics. These efforts strengthen the supply of high-quality data and support sustainable growth across the region.

Elaine S. Tan
Director, Data Division
Economic Research and Development Impact Department
Asian Development Bank

INTRODUCTION

1

Economic variables often display systematic and recurring within-year patterns influenced by various factors, such as weather conditions, institutional practices, and social customs. These seasonal effects can dominate short-term fluctuations in the data, making it difficult to distinguish between genuine economic signals and predictable calendar-related movements.[1] For example, retail sales may peak every December due to holiday shopping, while agricultural output may fluctuate with planting and harvest cycles. Without adjusting for such patterns, analysts risk misinterpreting seasonal highs and lows as meaningful economic changes.

Seasonal adjustment addresses this challenge by removing predictable seasonal and calendar effects from time series.[2] By doing so, it allows underlying non-seasonal movements, such as long-term growth trends, cyclical variations, and irregular shocks, to be more clearly identified. This is particularly important for timely detection of turning points in macroeconomic indicators such as quarterly national accounts, industrial production, or labor market statistics. For policymakers, researchers, and businesses alike, seasonally adjusted series provide a clearer view of the economy's true dynamics, supporting better decision-making.

Over the past decades, several tools for seasonal adjustment have become widely used. The most prominent are X-13ARIMA-SEATS, developed and maintained by the United States (US) Census Bureau,[3] and TRAMO/SEATS, originally created at the Bank of Spain.[4] Both approaches rely on model-based decomposition of time series and are recognized as international standards. More recently, these methods have been integrated into JDemetra+, an open-source platform jointly developed by the National Bank of Belgium, the Deutsche Bundesbank, and Eurostat. On 2 February 2015, JDemetra+ was officially recommended by the European Statistical System (ESS) and the European System of Central Banks as the reference software for seasonal and calendar adjustment of official statistics.

The purpose of this guide is to provide a practical introduction to seasonal adjustment using JDemetra+, with a focus on its application in the production of official statistics. The guide explains how to prepare time series data, define calendars and user regressors, apply both single and multiple seasonal adjustment processes, analyze outputs, refresh and export results, and communicate findings effectively to end users. By combining step-by-step instructions with applied examples, it aims to support statisticians and analysts in producing high-quality, transparent, and reproducible seasonal adjustment results that comply with international best-practice standards and strengthen the analytical foundations of official statistics within national statistical systems.

---

1 Eurostat. 2018. *Handbook on Seasonal Adjustment*. 2018 Edition. Luxembourg: Publications Office of the European Union. https://ec.europa.eu/eurostat/documents/3859598/8939616/KS-GQ-18-001-EN-N.pdf.

2 S. Hillmer and G. C. Tiao. 1982. An ARIMA-Model-Based Approach to Seasonal Adjustment. *Journal of the American Statistical Association*. 77 (377). pp. 63–70; A. Maravall and D. Pierce. 1987. A Prototypical Seasonal Adjustment Model. *Journal of Time Series Analysis*. 8 (2). pp. 177–193.

3 D. F. Findley, B. C. Monsell, W. R. Bell, M. C. Otto, and B.-C. Chen. 1998. New Capabilities and Methods of the X-12-ARIMA Seasonal-Adjustment Program. *Journal of Business & Economic Statistics*. 16 (2). pp. 127–152; B. C. Monsell. 2009. Update on the Development of X-13ARIMA-SEATS. *Proceedings of the Joint Statistical Meetings: American Statistical Association*. 13: 1–10.

4 V. Gómez and A. Maravall. 1995. *Programs TRAMO and SEATS*. Florence: European University Institute.

OVERVIEW OF JDEMETRA+ 2

JDemetra+ is an open-source application that was jointly created by the National Bank of Belgium, the Deutsche Bundesbank, and Eurostat. The software was designed to offer a standardized environment for carrying out seasonal and calendar adjustment of official statistics, ensuring that the process is transparent, reproducible, and consistent across institutions and countries. The importance of JDemetra+ was recognized formally on 2 February 2015, when it was officially recommended to members of the ESS and the European System of Central Banks as the reference tool for seasonal adjustment. Its development is also linked to the implementation of the ESS Guidelines on Seasonal Adjustment, most recently updated in 2024.[5]

At its core, JDemetra+ combines the two main approaches used worldwide for seasonal adjustment: TRAMO/SEATS, created at the Bank of Spain, and X-13ARIMA-SEATS, maintained by the US Census Bureau. These methods are considered the global standards for separating time series into trend, seasonal, and irregular movements. JDemetra+ brings these methods together within a modern, production-ready environment that makes advanced statistical modeling accessible to both specialists and practitioners. The software provides a comprehensive set of tools to guide users through the adjustment process, including RegARIMA modeling for pre-adjustment, residual analysis for assessing model adequacy, seasonality tests, spectral and sliding-span analyses, and a full range of quality diagnostics to ensure robust results.

From a technical perspective, JDemetra+ is designed as a suite of modular Java components that are both reusable and extensible, allowing users and developers to adapt the software to a wide variety of institutional needs. Although its interface resembles that of its predecessor Demetra+, JDemetra+ represents a significant leap forward in both scope and flexibility. Beyond seasonal adjustment, it offers a collection of open Java libraries that can handle large-scale datasets, support non-standard adjustment strategies, and enable the development of advanced research modules. Its extended functionality includes temporal disaggregation, benchmarking, business cycle analysis, and related applications, making it a comprehensive toolkit for time series analysis rather than a narrowly defined adjustment program. The software is also designed to work in combination with other environments, such as the R interface (RJDemetra) or the command-line tool known as the Cruncher, which facilitate automation and batch processing.

The first requirement for running JDemetra+ is the Java Runtime Environment, version 8 or higher, which can be freely downloaded from the Oracle website. Once Java is installed, JDemetra+ can be obtained from the official GitHub repository, where both stable releases (highlighted in green) and pre-releases (highlighted in red) are available. For most users, the stable release is recommended, as it is more suitable for regular production work. The program can be installed in two ways.

An installer is available for different operating systems (Windows, Linux, macOS, Solaris), or alternatively, a portable zip package can be selected, which requires no formal installation. After the zip package is downloaded, it can be extracted to any directory. The executable files are located in the "bin" folder of the extracted package. On Windows systems, the application is launched by double-clicking nbdemetra.exe for 32-bit environments or nbdemetra64.exe for 64-bit environments. Once these steps are completed, the program is ready for use, and time series data can be imported for seasonal adjustment.

[5] Eurostat. 2024. *ESS Guidelines on Seasonal Adjustment – 2024 Edition.* Luxembourg: Publications Office of the European Union.

PREPARATION OF TIME SERIES 3

3.1 Organizing Source Data

When JDemetra+ is launched, the main window opens with two panels on the left-hand side: the **Workspace** panel and the **Providers** panel (Figure 3.1.1).

Figure 3.1.1: JDemetra+ Default Window

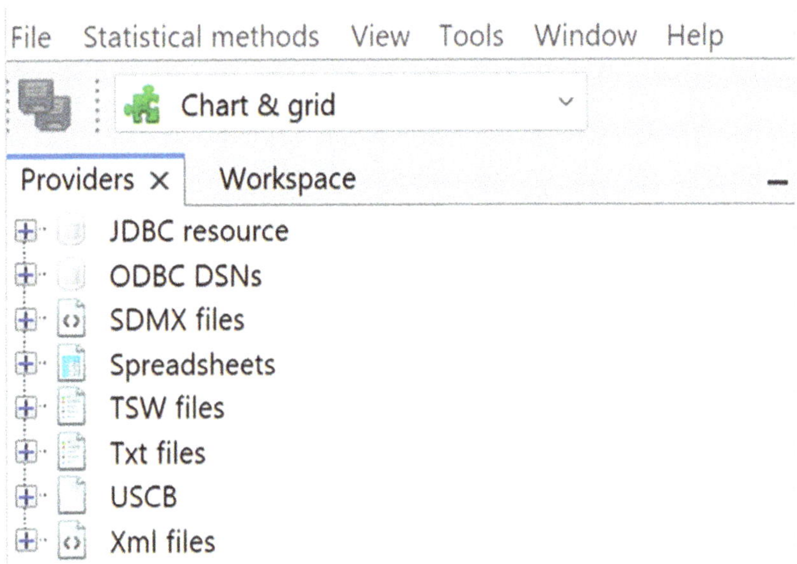

Source: Screenshot by the Asian Development Bank based on JDemetra+ 2.2.2.

The workspace is a central feature of JDemetra+, allowing users to store and organize their work in a coherent and structured way. By default, each workspace already includes predefined modeling and seasonal adjustment specifications, together with a basic calendar. Users can expand this by saving their own modeling and adjustment specifications, results of time series modeling and seasonal adjustment, user-defined calendars, and additional regression variables.

The **Providers** panel lists the available data sources and arranges imported series by provider. Out of the box, JDemetra+ supports common sources such as Java Database Connectivity connections, Excel spreadsheets and text files, X-13 files from the US Census Bureau, TRAMO/SEATS for Windows, as well as SDMX and XML files. Standard relational databases such as Oracle, SQLServer, DB2, or MySQL are also accessible through Java Database Connectivity. Users can extend the list of providers further by installing custom plugins.

Table 3.1.1: Structure of Source Data for JDemetra+

	Mar 2020	Jun 2020	Sep 2020	Dec 2020	Mar 2021
GDP	1,263,058.8	1,274,458.7	1,743,630.8	1,900,754.3	1,278,566.6
Total final consumption expenditure	1,239,881.9	1,221,110.3	1,518,017.7	1,602,939.1	1,239,140.7
Household final consumption expenditure	1,084,253.2	1,017,412.6	1,274,882.1	1,245,120.6	1,052,496.3
Final consumption expenditure of non-profits serving households	5,311.3	6,238.4	12,237.7	11,147.7	5,823.3
Government final consumption expenditure	150,317.4	197,459.3	230,897.9	346,670.8	180,821.1
Gross capital formation					
Gross fixed capital formation					
Changes in inventories	24,965.5	9,510.1	82,005.4	351.5	-27,091.3
Net exports of goods and services	-143,842.5	-115,590.3	-157,655.6	-198,557.0	-88,663.2
Exports	494,002.8	384,072.5	476,449.9	485,517.3	462,994.1
Imports	637,845.3	499,662.8	634,105.5	684,074.3	551,657.3
	Jun 2021	**Sep 2021**	**Dec 2021**	**Mar 2022**	**Jun 2022**
GDP	1,576,104.4	1,907,325.2	2,220,966.3	1,484,608.0	1,898,759.5
Total final consumption expenditure	1,424,541.1	1,619,470.6	1,848,146.5	1,389,894.0	1,686,173.0
Household final consumption expenditure	1,191,800.5	1,296,145.6	1,451,258.0	1,206,275.1	1,425,348.7
Final consumption expenditure of non-profits serving households	9,247.4	9,656.3	12,045.0	5,716.2	9,832.0
Government final consumption expenditure	223,493.2	313,668.7	384,843.5	177,902.7	250,992.3
Gross capital formation					
Gross fixed capital formation					
Changes in inventories	54,176.4	69,930.5	72,645.2	49,444.2	71,372.6
Net exports of goods and services	-99,016.8	-123,191.4	-282,317.6	-146,180.3	-121,561.7
Exports	629,502.4	686,849.5	688,152.2	623,431.8	782,448.1
Imports	728,519.2	810,040.9	970,469.8	769,612.1	904,009.8

GDP = gross domestic product.
Source: Statistical Committee of the Republic of Armenia.

When working with spreadsheets, the data must follow a specific structure, as illustrated in Table 3.1.1. Dates should be placed in the first column in a format recognized by Excel. For monthly data, the first day of each month can be used (for example, 1.1.2000, 1.2.2000, etc.), while for quarterly data the first day of each quarter is required (1.1.2000, 1.4.2000, 1.7.2000, etc.), ensuring that 3-month intervals are respected. The first row should contain the names of the variables, and the time series themselves must be entered as numbers. The top-left cell (A1) should be left empty. Any missing values must be manually coded as −99999; JDemetra+ will then estimate these using the additive outlier approach developed by Gómez, Maravall, and Peña.[6] A transposed format, with variable names in the first column and dates in the first row, is equally accepted.

6 V. Gómez, A. Maravall, and D. Peña. 1999. Missing Observations in ARIMA Models: Skipping Strategy versus Additive Outlier Approach. *Journal of Econometrics.* 88 (2). pp. 341–363.

3.2 Importing Data

Once the Excel source file meets the required formatting rules, it can be imported into JDemetra+ using a direct or an indirect procedure.

The direct import method works through the **Providers** tab (Figure 3.2.1) and is generally the recommended approach, particularly for repeated or official production processes. One of its advantages is that it enables dynamic updating: when a user needs to rerun an existing seasonal adjustment with a dataset that has been refreshed, JDemetra+ can automatically re-read the updated file and recalculate the seasonal adjustment results accordingly. To import a file directly, the user should right-click on **Spreadsheet** in the **Providers** tab and choose **Open**. In the dialogue box that appears, the file location can be set by clicking the "..." button next to **Spreadsheet** file under **Source** and then selecting the desired Excel file from the local folder. Once the file is loaded, right-clicking on it within the **Providers** tab allows further actions such as editing, renaming, or closing the file.

The indirect import method relies on copying and pasting data from the source file into JDemetra+. To do this, the user selects the entire dataset in Excel, including the column with the dates and the row with the variable names, and copies it. In JDemetra+, the user then navigates to **Tools** → **Container** → **Grid**, opens the **Grid** panel, and pastes the data inside. This method is simple and fast, making it useful for exploratory work or one-time seasonal adjustment exercises. However, since the copied dataset is not linked to the original file, dynamic refreshing is not possible.

Figure 3.2.1: Importing Data to JDemetra+

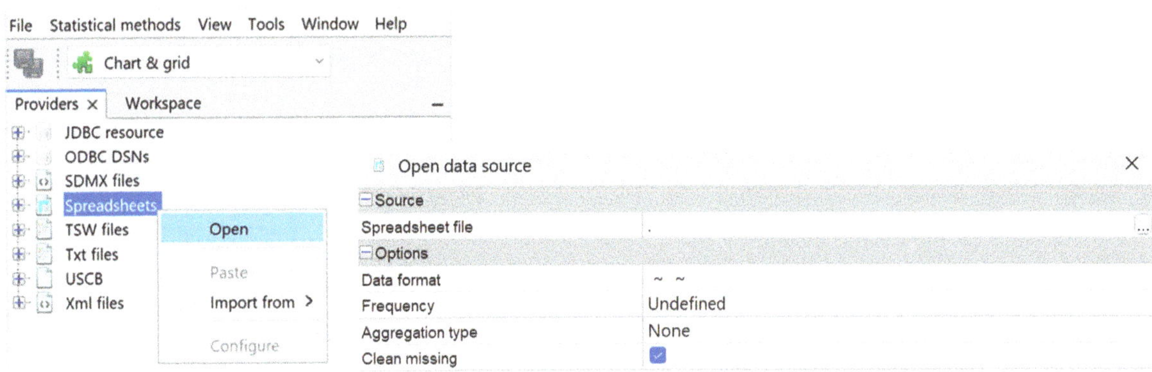

Source: Screenshot by the Asian Development Bank based on JDemetra+ 2.2.2.

3.3 Checking the Original Series

Before applying seasonal adjustment, it is essential to carefully examine the raw time series, since the reliability of the adjustment largely depends on the quality of the input data. Analysts should verify that the series is sufficiently long, complete, and consistent, and that its structure is suitable for decomposition into trend, seasonal, and irregular components. Visual inspection is particularly valuable for spotting potential issues such as missing values, outliers, abrupt level shifts, or changes in seasonal behavior.

Identifying these features in advance helps the analyst select an appropriate decomposition scheme and ensures that the model's assumptions remain valid. Recording these checks in a clear and systematic way also strengthens the robustness of the adjustment and provides transparency for colleagues and end-users who depend on the results.

In JDemetra+, the properties of a time series can be reviewed through **Tools → Container → List**. After opening the list panel, the user can simply drag the desired series from the Providers panel into it. To illustrate, consider quarterly national accounts by expenditure components for a hypothetical country. In the list view (Figure 3.3.1), the start and end dates of each series, the total number of observations, and a small line graph summarizing the data are displayed. If a component contains no information, JDemetra+ shows the warning "No data available." In our example, the available series consist of 38 quarterly observations, which, according to the ESS guidelines on seasonal adjustment, is sufficient for reliable seasonal adjustment using either model-based or moving average methods. By contrast, series shorter than 3 years cannot be adjusted with standard approaches, and those spanning only 3 to 6 years often lead to unstable results, particularly in calendar adjustment. For series shorter than 7 years, the ESS guidelines recommend close checking of parameter settings and, where possible, extending the data with reliable back-calculated observations.

Figure 3.3.1: List Panel Display of Variables

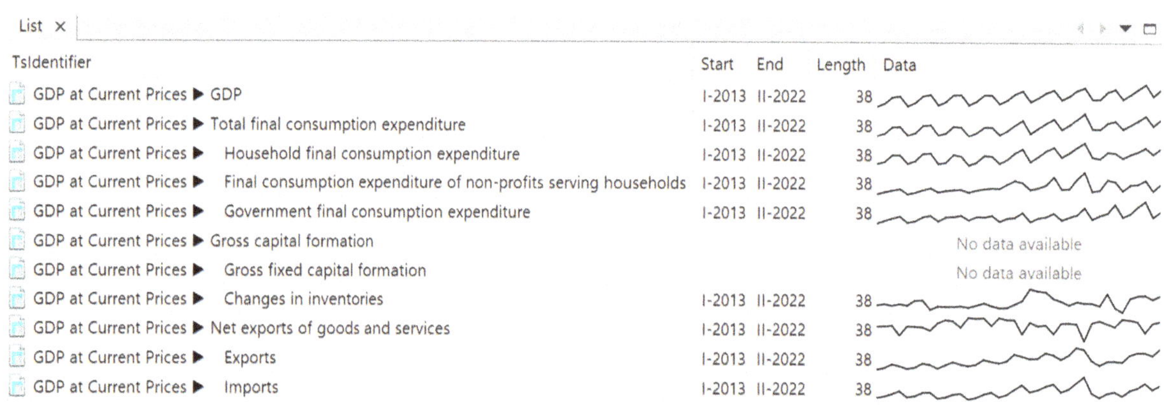

GDP = gross domestic product.
Source: Screenshot by the Asian Development Bank based on JDemetra+ 2.2.2.

Other useful entry points for exploring a time series in JDemetra+ are the **Chart**, **Growth Chart**, and **Spectral Analysis** options. The standard **Chart** provides a straightforward visual of the raw data, helping users detect patterns such as seasonality, trends, or sudden breaks. The **Growth Chart** focuses on growth rates, which makes it easier to identify turning points in dynamics that may not be immediately visible in the level series. **Spectral Analysis** offers a more advanced view by decomposing the series into its frequency components, allowing the analyst to confirm the presence of cyclical or seasonal behavior and to detect irregular fluctuations. Together, these tools complement the **List** view by giving both a descriptive and diagnostic perspective on the data, ensuring that the analyst develops a solid understanding of the series before proceeding with seasonal adjustment.

DEFINING CALENDARS AND EXTERNAL VARIABLES 4

Calendar adjustment is an essential part of seasonal adjustment and is typically carried out during the pre-treatment stage of the process. Its main objective is to correct for the impact of calendar-related effects, such as variations in the number of working days, leap years, and holidays, so that the subsequent decomposition can accurately identify and estimate the genuine seasonal and trend components. Without correcting for these effects, apparent "seasonal" movements may simply reflect differences in the calendar structure rather than true seasonal behavior, leading to biased estimates and misinterpretation of short-term economic dynamics.[7] In this section, we describe how calendar effects can be identified and adjusted for within the seasonal adjustment process, including the creation of a national calendar and the incorporation of calendar variables in JDemetra+.

4.1 Creating a National Calendar

Once the source data are prepared, the next step is to define a national calendar, which allows to improve the estimation of calendar effects in seasonal adjustment. JDemetra+ offers a dedicated **Calendars** module that allows users to generate calendar variables such as trading day (six variables plus a leap year effect), working day (one variable plus leap year), and country- or sector-specific holidays. These holidays may be either fixed (e.g., 1 January) or moving (e.g., Easter).

To define a national calendar in JDemetra+, the option **Add Calendar → National** should be selected from the context menu of the **Calendars** item in the **Workspace** window. A window will then be opened in which the name of the calendar can be entered in the **Name** field. Holidays are added by expanding the "+" menu, while an existing holiday can be removed by selecting it in the list and pressing the "–" button (Figure 4.1.1).

Within this window, four types of holidays can be specified:

- **Fixed** – a holiday that occurs on the same calendar date every year (e.g., 1 January).
- **Easter-related** – a holiday whose date is determined in relation to Easter.
- **Fixed Week** – a holiday that consistently falls on a particular weekday in a given month (e.g., the second Monday in October).
- **Special Day** – a holiday selected from a predefined list that includes widely used fixed and moving holidays.

When a national calendar is defined in JDemetra+, the corresponding regression variables also include a mean effect, meaning the average contribution of the calendar that is independent of seasonal variation. In the usual decomposition of a time series, this mean effect is attributed to the trend-cycle component, while the actual calendar effect reflects only the deviations that cannot be explained by either trend or seasonality. Since a year contains on average 365.25 days, the monthly means are approximately 21.74 weekdays, 8.70 weekend days, and 30.44 total days. To avoid misinterpreting these averages as seasonal fluctuations, the long-term mean correction option should always be selected when defining a national calendar.

[7] E. B. Dagum and S. Bianconcini. 2016. *Seasonal Adjustment Methods and Real-Time Trend-Cycle Estimation.* Cham: Springer.

Figure 4.1.1: Adding a National Calendar

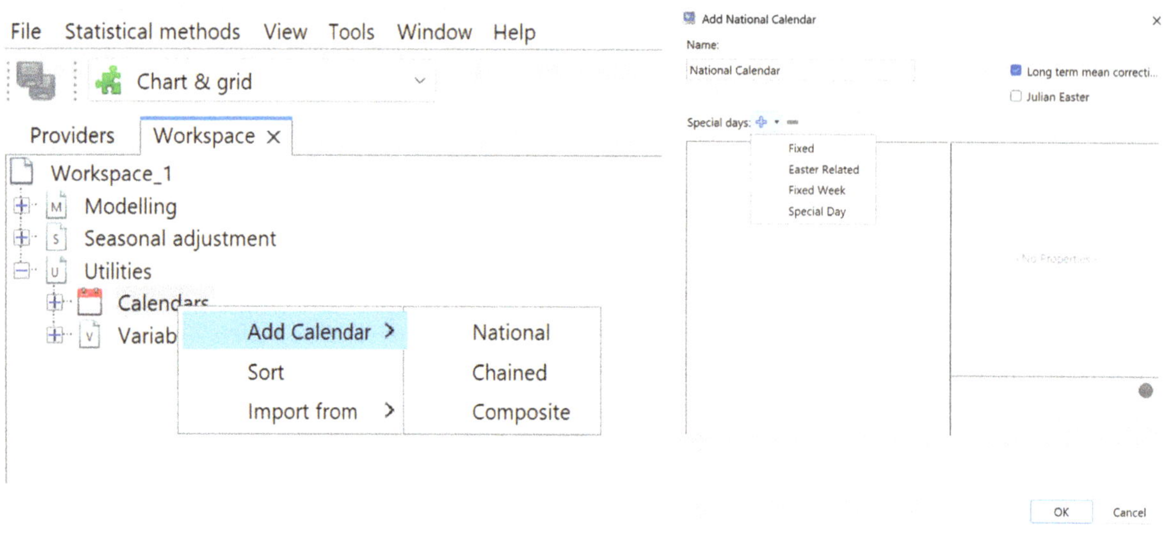

Source: Screenshot by the Asian Development Bank based on JDemetra+ 2.2.2.

4.2 Special Days

JDemetra+ includes a set of 17 predefined holidays, reflecting the most common observances across Europe (Table 4.2.1). However, since holiday practices differ from country to country, the default list should always be reviewed and adjusted to ensure the correct dates are used.

When the **Special Days** option is activated, the program automatically inserts Christmas into the list of holidays. This default choice can be modified in the settings panel on the right, where adjustments are confirmed by clicking "OK". The editable parameters are as follows:

- **Start** – defines the first year in which the holiday is considered. By default, this field is empty, which corresponds to the start of the calendar. If a custom date is required, it should be entered in the format yyyy-mm-dd.
- **End** – specifies the final year in which the holiday is valid. As with the start field, the default is an empty cell, meaning the holiday applies until the end of the calendar. User-defined dates must also follow the yyyy-mm-dd format.
- **Weight** – determines how strongly the holiday influences the time series. The default value is 1, which is interpreted as equivalent to the impact of a Sunday. If evidence suggests that the holiday has a smaller effect, a weight between 0 and 1 may be chosen. The choice of weight should ideally be guided by expert judgment or empirical analysis.
- **Day event** – provides access to the list of predefined holidays that can be selected.
- **Offset** – indicates the timing of the holiday relative to the predefined holiday selected. The default value is 1. A positive offset places the event after the chosen holiday, while a negative offset assigns it to the days before.

Table 4.2.1: Default Special Holidays in JDemetra+

Holiday	Definition
New Year	A fixed holiday falling on 1 January
Shrove Monday	A moving holiday falling on the Monday before Ash Wednesday
Shrove Tuesday	A moving holiday falling on the day before Ash Wednesday
Ash Wednesday	A moving holiday occurring 46 days before Easter
Easter	A moving holiday varying between 22 March and 25 April, always on a Sunday
Maundy Thursday	A moving holiday falling on the Thursday before Easter, the fifth day of Holy Week
Good Friday	A moving holiday on the Friday in Easter week
Easter Monday	A moving holiday falling on the Monday after Easter Sunday
Ascension Day	A moving holiday on the Thursday, the 40th day after Easter
Pentecost	A moving holiday celebrated 50 days after Easter Sunday, 10 days after Ascension Thursday
Corpus Christi	A moving holiday two months after Maundy Thursday
Whit Monday	A moving holiday celebrated on the Monday after Pentecost
May Day	A fixed holiday on 1 May, usually a public holiday and spring festival
Halloween	A fixed holiday on 31 October
Assumption	A fixed holiday on 15 August
All Saints Day	A fixed holiday on 1 November (Western Christianity) or the first Sunday after Pentecost (Eastern Christianity)
Christmas	A fixed holiday on 25 December in many countries

Source: S. Grudkowska. 2015. *JDemetra+ User Guide*. Warsaw: National Bank of Poland.

4.3 Fixed Days

Most countries include fixed-date holidays in their national holiday calendars. These are holidays that occur on the same day and month every year, such as 1 January (New Year's Day). At first glance, one might assume that their impact is automatically absorbed by the seasonal component of a time series. However, this is not always the case, because fixed holidays may fall either on a weekday or on a weekend, and this irregular shifting can distort the seasonal pattern. For this reason, it is generally recommended to include fixed holidays explicitly as calendar regressors in JDemetra+.

In practice, many fixed holidays are not part of the program's predefined set. These can easily be added as fixed holidays by selecting **Fixed** from the **Special Days** menu and entering the required details (Figure 4.3.1).

The initial holiday shown by default can be modified through the settings panel on the right-hand side. Several parameters can be adjusted here. The **Start** field determines the first year in which the holiday is valid; if left blank, it is assumed to begin at the start of the calendar. Dates must be entered in the format yyyy-mm-dd.

Similarly, the **End** field indicates the last year for which the holiday applies, with an empty cell corresponding to the end of the calendar. The **Weight** parameter defines the strength of the holiday's effect on the series. Its default value is 1, which means the holiday is treated as equivalent to a Sunday. If the impact is expected to be smaller, any value between 0 and 1 can be assigned, though such adjustments should be based on expert judgement or empirical studies. Finally, the **Day** and **Month** fields specify the exact calendar date on which the fixed holiday occurs each year.

Figure 4.3.1: Adding a Fixed Holiday

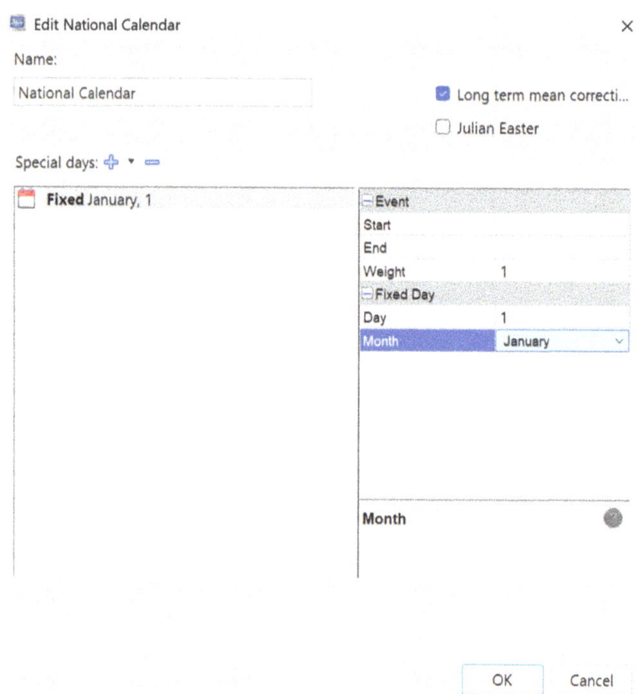

Source: Screenshot by the Asian Development Bank based on JDemetra+ 2.2.2.

4.4 Easter-Related Holidays

The exact date of Easter varies each year, and the timing as well as the number of associated holidays can differ across countries. Because these holidays are linked to Easter but not always celebrated in the same way, JDemetra+ includes a specific option called **Easter-related Days** that allows users to define them flexibly (Figure 4.4.1). This option can be selected from the **Special Days** list, after which the relationship between Easter and the additional holiday is determined by specifying an offset in days.

Easter-related holidays in JDemetra+ are defined in much the same way as fixed holidays: the user can specify the **Start** and **End** years of validity and, if necessary, adjust the **Weight** of the effect. The key distinction is that these holidays are not tied to a fixed calendar date but instead move with the timing of Easter. To account for this, an additional parameter, the **Offset**, is used to position the holiday in relation to Easter. By default, the setting is +1 (Easter Monday), but the offset can take positive or negative values, for example, −2 for Good Friday.

Figure 4.4.1: Adding an Easter-Related Holiday

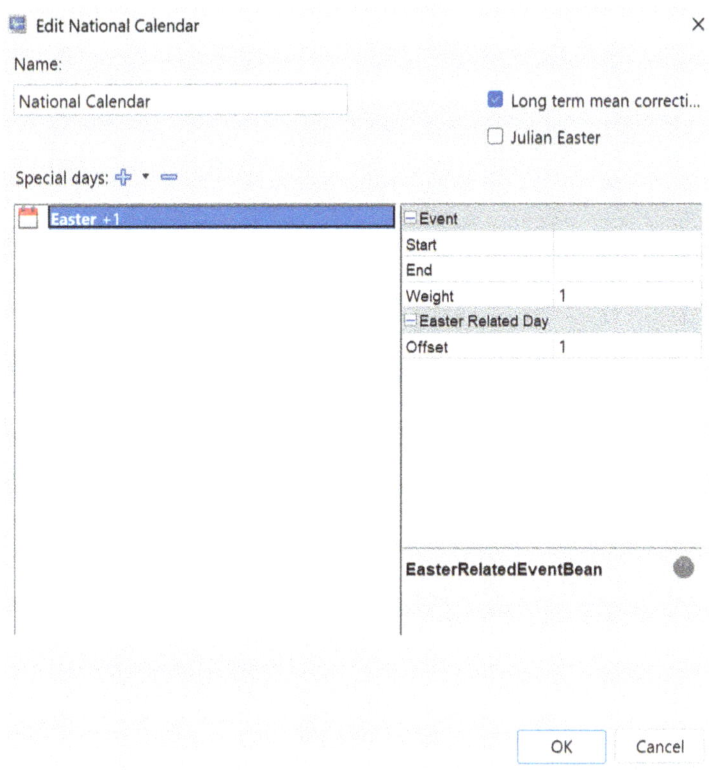

Source: Screenshot by the Asian Development Bank based on JDemetra+ 2.2.2.

4.5 Fixed Week

The **Fixed Week** option is designed for holidays that always fall on the same weekday within a specific week of a month. A typical example is Labor Day in Canada, which is observed on the first Monday of September each year. To create such an event, the **Fixed Week** item should be selected from the **Special Days** list (Figure 4.5.1).

The **Fixed Week** option in JDemetra+ relies on the same core parameters as other holiday types: the **Start** and **End** dates, which define the period of validity, and the **Weight**, which measures the strength of the holiday's effect on the series (defaulting to 1, equivalent to a Sunday, but adjustable between 0 and 1 if the impact is weaker). What makes this option distinct is that the holiday is not tied to a specific calendar date or to Easter but instead to a recurring weekday in a given month. For this reason, three additional fields are required: the **Day of the Week** (e.g., Monday), the **Month** (e.g., September), and the **Week number** within that month (ranging from 1 to 5).

Figure 4.5.1: Adding Fixed Week to the National Calendar

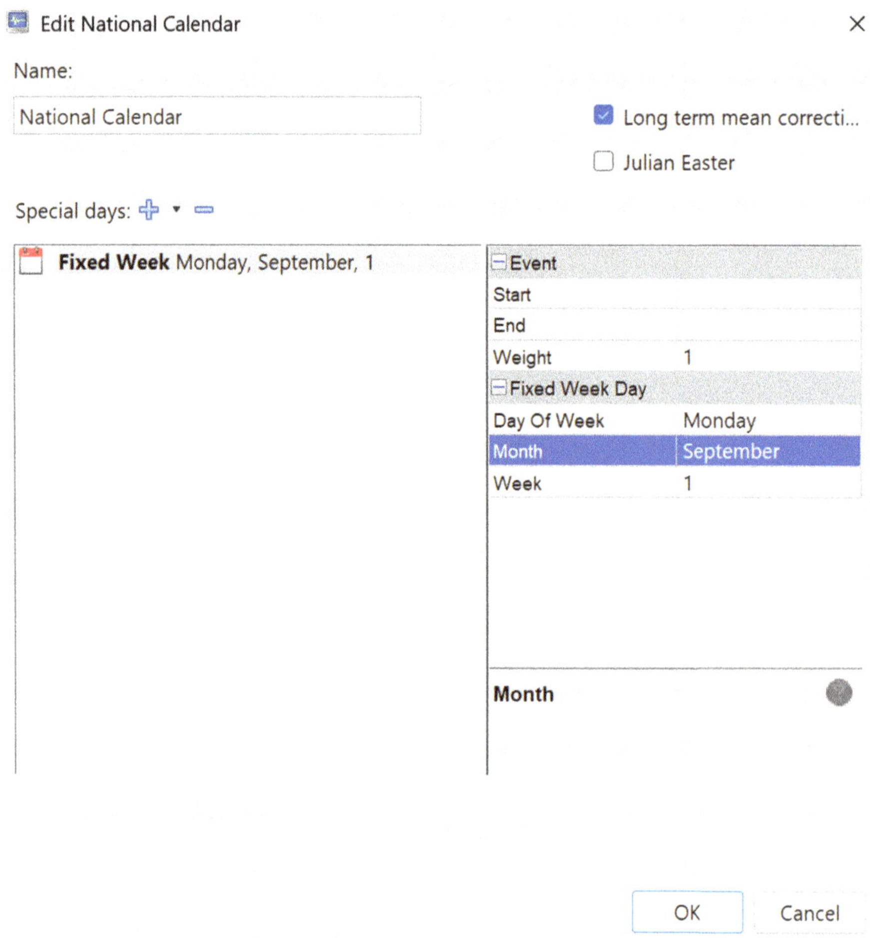

Source: Screenshot by the Asian Development Bank based on JDemetra+ 2.2.2.

4.6 External User-Defined Variables

In some cases, holidays cannot be created directly within JDemetra+ because their timing follows a religious or lunar calendar. For instance, in Indonesia, the dates of Ramadan and Eid al-Adha (Feast of Sacrifice) shift each year relative to the Gregorian calendar. To capture their effect, a user-defined calendar variable must be prepared externally in Excel and then imported into JDemetra+.

The first step is to determine the exact dates of the holidays for the analysis period. Table 4.6.1 shows the design of the Excel file, where each column represents either a moving holiday (Ramadan, Eid al-Adha) or an official fixed holiday (e.g., Independence Day, New Year). Saturdays and Sundays are also included in the file to check whether moving holidays overlap with regular non-working days. A value of 1 is assigned for each day when the holiday occurs, and 0 otherwise. Two versions of the variable can be constructed: one that excludes Saturdays, Sundays, and official holidays, and one that excludes only Sundays and official holidays.

Table 4.6.1: Structure of an Excel File for External User-Defined Variables

Date	Sacrifice	Ramadan	New Year	Saturday	Sunday	**Sacrifice** (Excluding Saturday, Sunday, official holidays)	**Sacrifice** (Excluding Sunday, official holidays)	**Ramadan** (Excluding Saturday, Sunday, official holidays)	**Ramadan** (Excluding Sunday, official holidays)
1.1.1974	0	0	1	0	0	0	0	0	0
2.1.1974	0	0	0	0	0	0	0	0	0
3.1.1974	0	0	0	0	0	0	0	0	0
4.1.1974	1	0	0	0	0	1	1	0	0
5.1.1974	1	0	0	1	0	0	0	0	0
6.1.1974	1	0	0	0	1	0	0	0	0
...	0	0	0	1	0	0	0	0	0

Source: Asian Development Bank illustration based on a hypothetical scenario.

Once the daily data are prepared, the second step is to aggregate them into monthly values (Table 4.6.2). The aggregation is straightforward: for each month, the values in the daily file (coded as 1 when a holiday occurs and 0 otherwise) are **summed** to give the total number of holiday days in that month. The result is a set of monthly regressors that show, for every month in the dataset, how many days of each holiday occurred.

Table 4.6.2: Aggregated Monthly Values

Date	**Sacrifice** (Excluding Saturday, Sunday, official holidays)	**Sacrifice** (Excluding Sunday, official holidays)	**Ramadan** (Excluding Saturday, Sunday, official holidays)	**Ramadan** (Excluding Sunday, official holidays)
1.1.1974	2	0	3	0
1.2.1974	0	0	0	0
1.3.1974	0	0	0	0
1.4.1974	0	0	0	0
1.5.1974	0	0	0	0
...	0	0	0	0

Source: Asian Development Bank illustration based on a hypothetical scenario.

The third step is to center the aggregated monthly variables (as shown in Table 4.6.3) to make them suitable for seasonal adjustment. JDemetra+ requires user-defined regressors to have a mean of zero over the entire sample period. Centering ensures that the regressors reflect only the short-term deviations caused by holidays, rather than introducing permanent shifts into the series. For instance, if Ramadan in April lasts 12 days on average but extends to 15 days in a particular year, the centered value for that month will be +3. Conversely, if it lasts only 10 days, the centered value will be –2.

Table 4.6.3: Mean-Adjusted Regressors

Date	Sacrifice (Excluding Saturday, Sunday, official holidays)	Average Values (Sacrifice)	Sacrifice Regressor (Excluding Sunday, official holidays)	Ramadan (Excluding Saturday, Sunday, official holidays)	Average Values (Sacrifice)	Ramadan Regressor (Excluding Saturday, Sunday, official holidays)
1.1.1974	2	0.2381	1.7619	0	0.50	-0.50
1.2.1974	0	0.2143	-0.2143	5	0.47	4.53
1.3.1974	0	0.2143	-0.2143	0	0.45	-0.45
1.4.1974	0	0.1905	-0.1905	0	0.40	-0.40
1.5.1974	0	0.1429	-0.1429	0	0.35	-0.35
...	0	0.2143	-0.2143	0	0.30	-0.30

Source: Asian Development Bank illustration based on a hypothetical scenario.

Once the external calendar variables have been prepared and centered, they need to be imported into JDemetra+ so that they can be used in the seasonal adjustment process. There are two main ways to do this:

1. Static import (manual copy or drag-and-drop)

A new placeholder variable must first be created inside JDemetra+. This is done by right-clicking on **Workspace → Utilities → Variables** and selecting **New**. A default item, usually labelled Vars-1, appears in the list. By double-clicking this entry, the user opens an empty container where the series can be inserted. At this stage, the data can either be copied directly from Excel and pasted into the container (using Ctrl+C and Ctrl+V), or dragged and dropped from an open spreadsheet window. This method creates a static version of the variable: once imported, it does not automatically update if the original Excel file is changed.

At this stage, it is good practice to verify whether the variable behaves as intended by checking its descriptive statistics and graphical display. In particular, the centered variable should oscillate around zero and reflect only the deviations from the long-term average number of holiday days. If the series shows a systematic upward or downward level, it usually means that centering has not been applied correctly. To make the variable functional within the seasonal adjustment framework, the specification of calendar effects must be switched to **UserDefined** in the **Specifications** menu (Figure 4.6.1). Once this option is selected, JDemetra+ will treat the imported series as the official calendar regressor for the adjustment. The user can then specify which of the external variables should be applied, either individually (e.g., only Ramadan or Sacrifice) or in combination (e.g., a total holiday variable).

Figure 4.6.1: Adding User-Defined Regressors

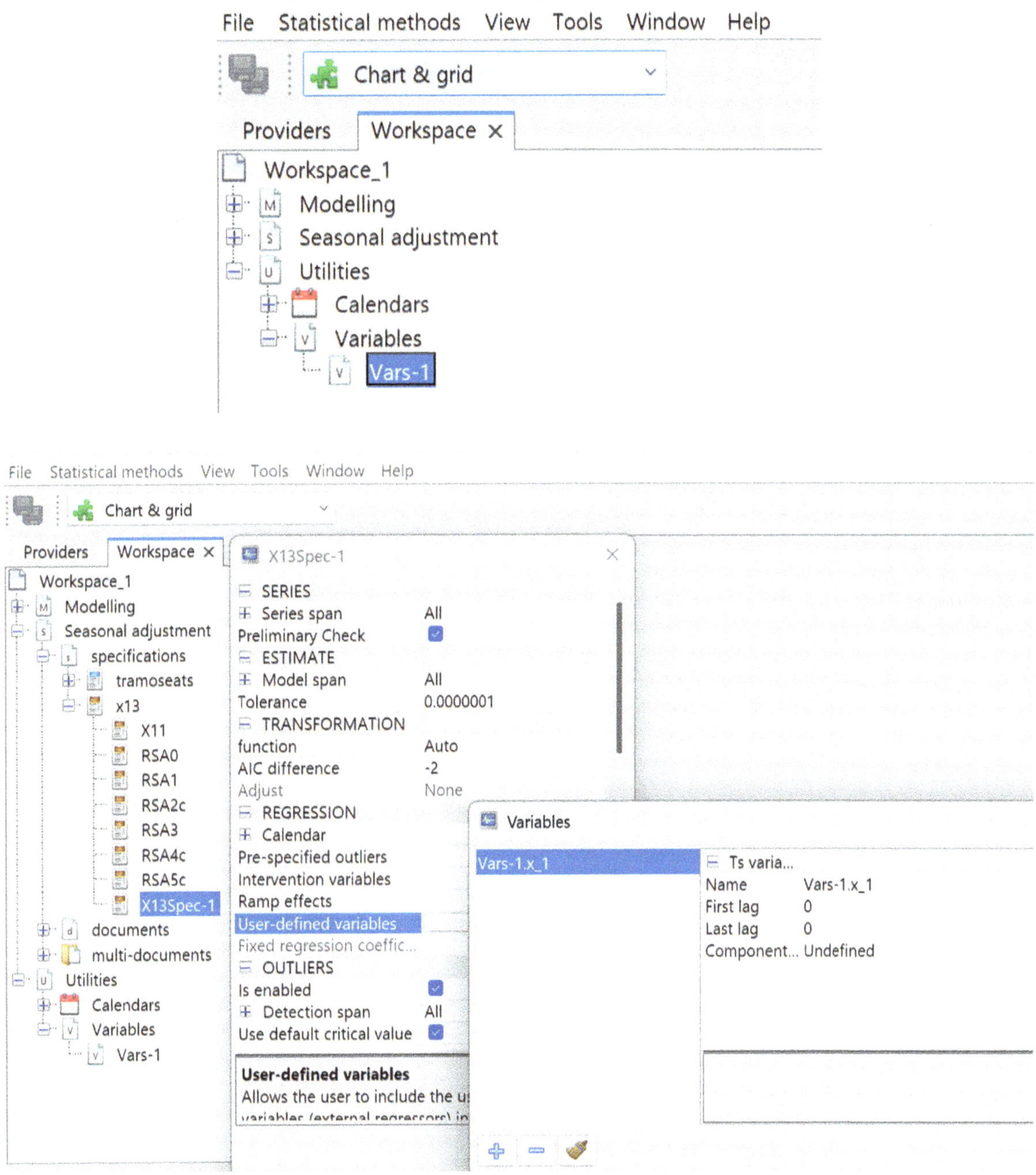

Source: Screenshot by the Asian Development Bank based on JDemetra+ 2.2.2.

Calendar regressors can also be generated in JDemetra+ using the TransReg plugin (**Tools → TransReg**). If **TransReg** is not listed under **Tools**, it must be first installed via **Tools → Plugins → Downloaded → Add Plugins**, selecting the previously downloaded plugin file (Figure 4.6.2). After installation, open **TransReg**, users can navigate to **Providers → Spreadsheet**, and drag the Excel sheet containing the monthly (or quarterly) aggregated values into the **TransReg** window (into "Drop data here"), then click the green button to proceed. The plugin will automatically generate the corresponding centered monthly (or quarterly) calendar regressors from the aggregated values. It is recommended to save the generated calendar regressors to a separate Excel sheet.

Figure 4.6.2: Installing TransReg Plugin

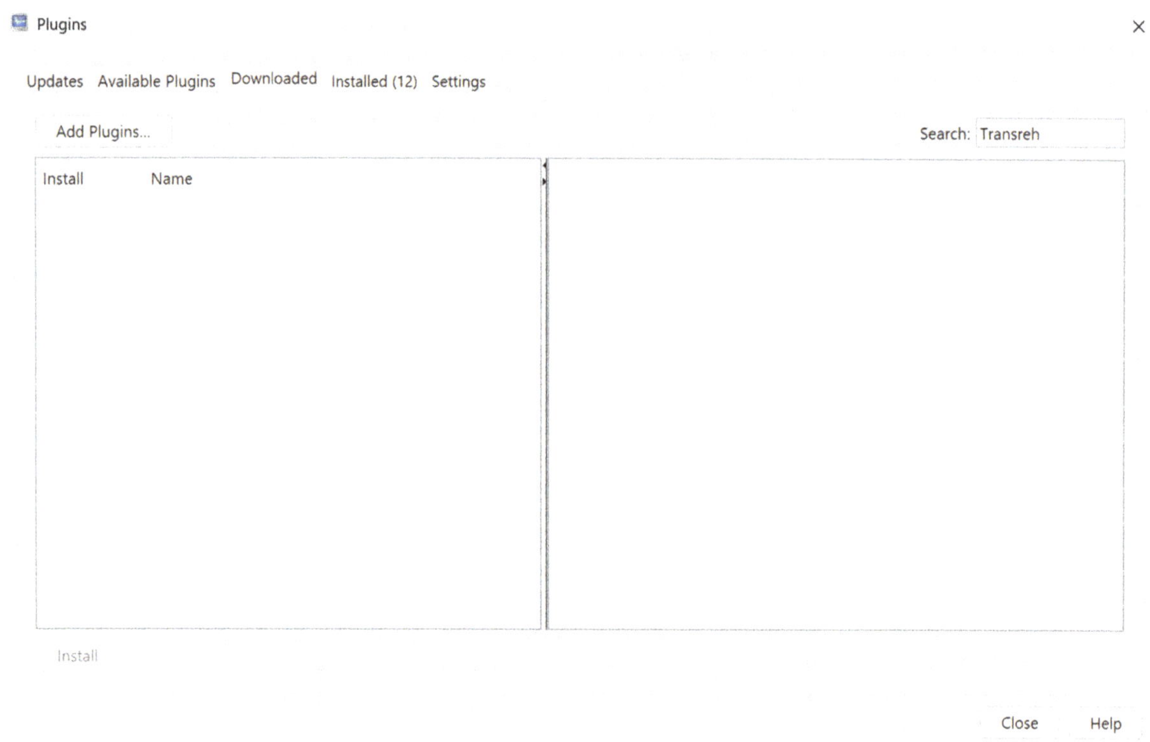

Source: Screenshot by the Asian Development Bank based on JDemetra+ 2.2.2.

2. Dynamic import (linking to Excel)

Instead of manually pasting the data, the Excel file can be linked to JDemetra+ through the **Providers** panel. This option allows the program to read the data directly from the source file, ensuring that the variable is refreshed automatically whenever the Excel file is updated. Dynamic imports are particularly useful in production environments, where the holiday variables need to be extended each year and recalculated without rebuilding the file from scratch.

SEASONAL ADJUSTMENT

5

5.1 Specifications and User-Regressors

Once the calendar effects have been defined and added, the next stage is seasonal adjustment. This process is not a single step, but a structured sequence of operations that work together to isolate the seasonal component from the raw data. These operations usually include pre-adjustments (such as correcting for outliers, working days, or moving holidays), decomposition of the series into trend, seasonal, and irregular parts, and finally a set of diagnostic checks that assess the quality and stability of the results.

In JDemetra+, the way these steps are carried out is governed by what is called a specification. A specification is essentially the blueprint that tells the program how to perform each part of the adjustment. It brings together all the key decisions: how to treat outliers, which the ARIMA model is used to estimate, whether to apply transformations, how to handle calendar regressors, and how to evaluate model adequacy through diagnostics. Users may either construct a specification manually, adjusting each parameter to reflect the characteristics of the series, or make use of the predefined templates supplied within the software. These templates cover the most common adjustment needs but can be modified to suit special data patterns or institutional conventions. All specifications are available under **Workspace → Seasonal Adjustment → Specification**. For X-13, JDemetra+ provides seven predefined specifications, each with default settings that aim to combine flexibility with ease of use (Table 5.1.1).

Table 5.1.1: Specification Options for X-13

Specification	Transformation	Leap Year	Working Days	Trading Days	Easter	Outliers	ARIMA Model
X11	No	No	No	No	No	No	None (pure X11 filters)
RSA0	No	No	No	No	No	No	(0,1,1)(0,1,1)
RSA1	Test	No	No	No	No	Test	(0,1,1)(0,1,1)
RSA2c	Test	Test	Test	No	Test	Test	(0,1,1)(0,1,1)
RSA3	Test	No	No	No	No	Test	AMI
RSA4c	Test	Test	Test	No	Test	Test	AMI
RSA5	Test	Test	No	Test	Test	Test	AMI

Source: JDemetra+ Reference Manual.

The predefined specifications in JDemetra+ for X-13ARIMA-SEATS offer users a set of ready-made options that balance simplicity, flexibility, and robustness. Each specification combines different settings for transformation, calendar adjustment, outlier detection, and ARIMA modeling, with the main predefined options including the following:

- **X11:** Relies exclusively on X11 filters, with no ARIMA modeling, no transformations, and no corrections for calendar or outlier effects. It is purely a filter-based method, useful when a simple decomposition is desired.
- **RSA0:** Does not include transformations, calendar effects, or outlier detection. The ARIMA model is fixed to the Airline model (0,1,1)(0,1,1). It represents a very basic model-based approach, assuming stable series behavior.
- **RSA1:** Adds a transformation test to select between additive or multiplicative decomposition. Outliers are identified automatically, but no calendar regressors are included. The ARIMA model is fixed to the Airline model.
- **RSA2c:** Extends RSA1 by introducing calendar regressors: leap-year correction, working days, and Easter. Outlier detection is automatic, and the ARIMA model remains fixed to the Airline model. This is a standard option for series where calendar variation is expected.
- **RSA3:** Similar to RSA1, but the ARIMA model is not fixed. Instead, automatic model identification (AMI) is applied to search for the best-fitting specification. It includes transformation tests and outlier detection but excludes calendar regressors.
- **RSA4c:** Combines the flexibility of RSA3 with the calendar corrections of RSA2c. It performs transformation tests, detects outliers, and applies automatic ARIMA model identification, along with leap-year, working days, and Easter effects. This makes it one of the most commonly recommended specifications.
- **RSA5:** The most comprehensive option, applying transformation tests, leap-year and trading-day corrections, Easter effects, outlier detection, and automatic ARIMA model identification. It is suitable for complex series where multiple sources of variation are likely.

In addition to the predefined options, JDemetra+ allows users to create their own tailored specifications. This is particularly useful when the default templates do not fully capture the characteristics of the series or when institutional requirements call for specific parameter settings. A new specification can be created by right-clicking on **X13** under **Workspace → Seasonal Adjustment → Specifications** and selecting **New**. The new specification will then appear in the **Workspace** panel, and by double-clicking it – or right-clicking and selecting **Open** – a dialogue box will open where its properties can be specified (Figure 5.1.1).

Figure 5.1.1: Layout of the Specification Dialogue Box

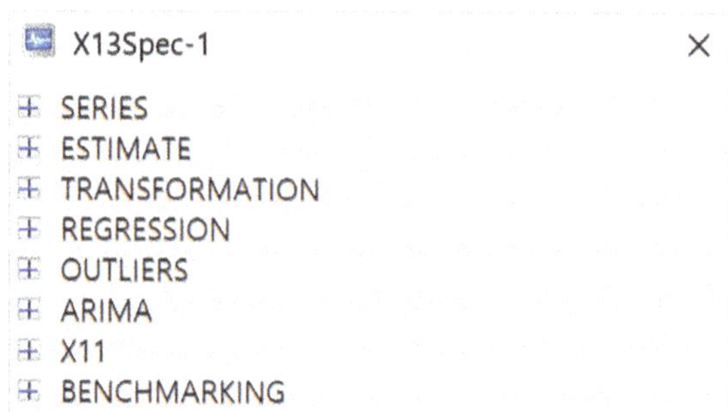

Source: Screenshot by the Asian Development Bank based on JDemetra+ 2.2.2.

Another way to create customized specifications in JDemetra+ is by duplicating an existing default template and then adjusting it. To do this, users can right-click on any of the default specifications found under **Workspace → Seasonal Adjustment → Specifications → X13** and choose **Clone** (Figure 5.1.2).
The duplicated specification will appear in the **Workspace** and can then be edited in exactly the same manner as a newly created specification.

Figure 5.1.2: Workflow for Custom Specification Setup

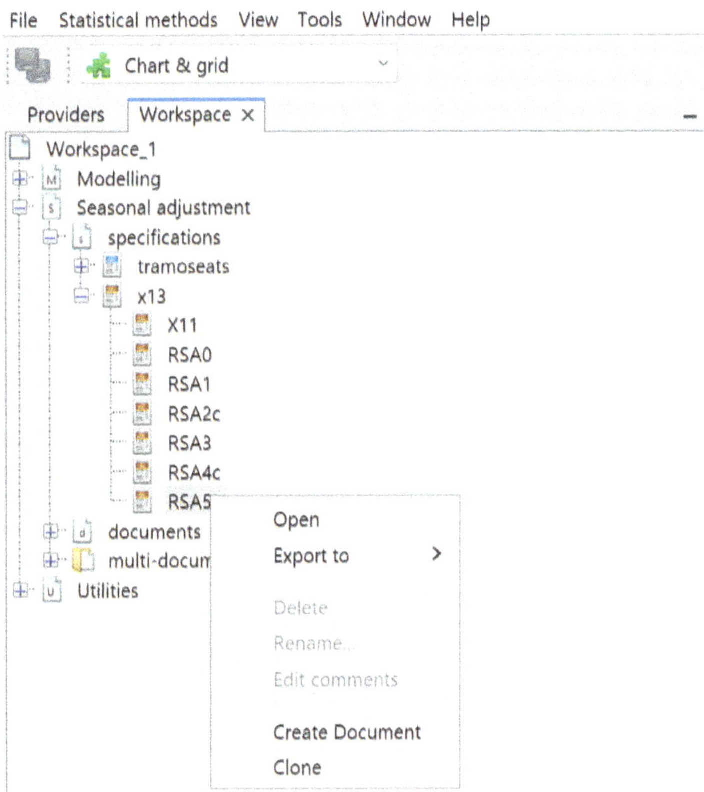

Source: Screenshot by the Asian Development Bank based on JDemetra+ 2.2.2.

Series

The first element of the specification dialogue is the **Series** panel (Figure 5.1.3), which contains two main options: **Series span** and **Preliminary check**. The **Series span** option allows the user to restrict the time interval of the data used for modeling or adjustment to the most relevant portion of the series. Depending on the setting, the user may choose to apply the procedure to the entire series, start from a particular date, end at a given date, define both a start and an end point, limit the adjustment to the most recent or earliest observations, or exclude a specified number of observations from the beginning and/or end of the series. The **Preliminary check** serves as a quality control mechanism. When this option is selected, the program automatically evaluates the input series and excludes those that do not meet minimum quality standards, such as series containing too many outliers, repeated identical values, or missing data beyond predefined thresholds. If the box is left unchecked, these thresholds are ignored and the program attempts to proceed with the adjustment whenever possible.

Figure 5.1.3: Series Panel

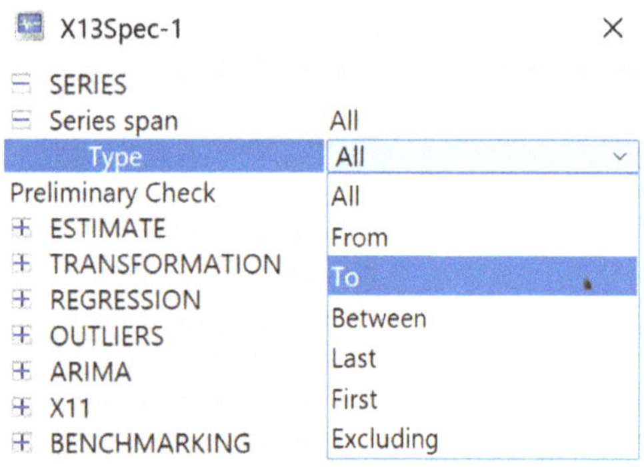

Source: Screenshot by the Asian Development Bank based on JDemetra+ 2.2.2.

Figure 5.1.4: Estimate Panel

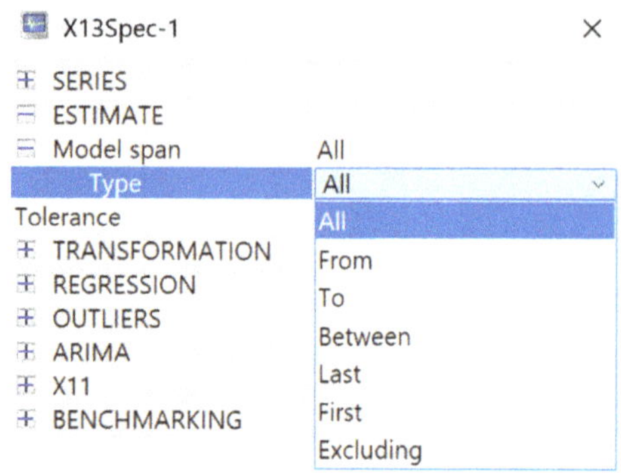

Source: Screenshot by the Asian Development Bank based on JDemetra+ 2.2.2.

Estimate

The next component of the specification dialogue is the **Estimate** panel, which defines the parameters of the RegARIMA model estimation (Figure 5.1.4). In most applications, the default settings in this section are sufficient and do not require modification.[8] An exception is the **Model Span**, which allows the user to restrict the portion of the data used for estimating the RegARIMA model. Importantly, even if the model is estimated on a limited span, the resulting specification is subsequently applied to the entire series when performing seasonal adjustment.

Another key parameter is **Tolerance**, which sets the convergence criterion for the nonlinear optimization procedure. During estimation, successive values of the log-likelihood function are compared, and the algorithm stops once the absolute changes fall below the specified tolerance level. This ensures numerical stability and prevents overfitting by terminating the iterations once a sufficiently precise solution has been reached.[9] The default value is 0.0000001, which provides a balance between accuracy and computational efficiency.

Transformation

The **Transformation** panel defines how the series is pre-processed before the RegARIMA model is estimated, and it consists of three main elements: **Function**, **AIC difference**, and **Adjust**. The **Function** setting offers three choices: **None** (no transformation), **Log** (logarithmic transformation of the series), and **Auto** (an automatic test that selects between a log or level specification).[10] When the **Auto** option is selected, the program also applies the **AIC difference** criterion, which sets the minimum improvement in the corrected Akaike Information Criterion (AICC) needed to prefer no transformation over a log

8 United States Census Bureau. 2023. *X-13ARIMA-SEATS Reference Manual*. Washington, DC: US Department of Commerce.

9 E. B. Dagum and S. Bianconcini. 2016. *Seasonal Adjustment Methods and Real-Time Trend-Cycle Estimation*. Cham: Springer; B. C. Monsell. 2009. Update on the Development of X-13ARIMA-SEATS. *Proceedings of the Joint Statistical Meetings: American Statistical Association*. 13: 1–10.

10 G. E. P. Box and D. R. Cox. 1964. An Analysis of Transformations. *Journal of the Royal Statistical Society: Series B (Methodological)*. 26 (2). pp. 211–243.

transformation; the default value is −2, and the option is only active under **Auto**. The final element, **Adjust**, applies proportional corrections for calendar effects and is available only when the Function is set to **Log**. Users can choose **LeapYear** (adjustment for February in leap years), **LengthOfPeriod** (adjustment for varying lengths of months or quarters), or **None** (no adjustment), with the default set to **None** (Figure 5.1.5).

Figure 5.1.5: Transformation Panel

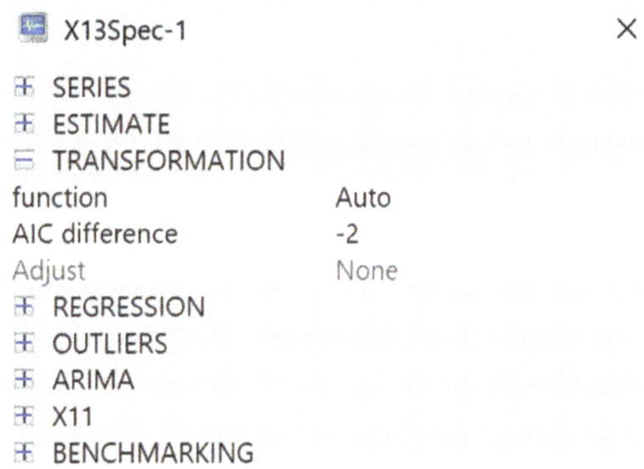

Source: Screenshot by the Asian Development Bank based on JDemetra+ 2.2.2.

Regression

The **Regression** panel captures deterministic effects in the series through predefined variables, structured into five categories: **Calendar, Pre-specified outliers, Intervention variables, Ramp effects, and User-defined variables** (Figure 5.1.6).

Figure 5.1.6: Regression Panel

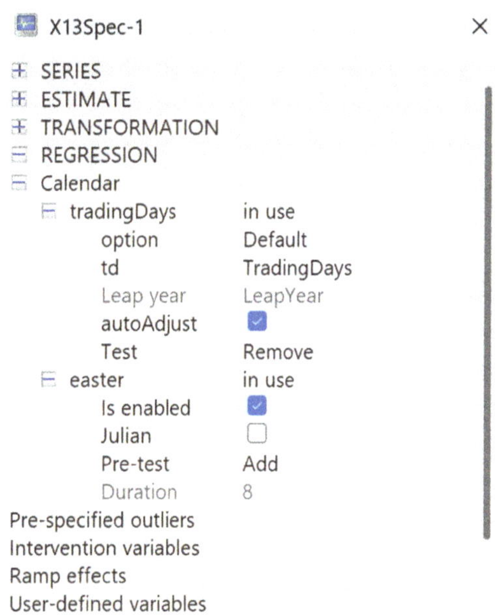

Source: Screenshot by the Asian Development Bank based on JDemetra+ 2.2.2.

The first of these, **Calendar**, is especially important, as it accounts for systematic variations in the data linked to the calendar, such as differences in the number of working or trading days, the presence of movable holidays like Easter, or leap-year adjustments. Correcting for these factors is essential to avoid distortions in seasonal adjustment and to ensure that the underlying trends are properly identified. The main calendar specifications are:

- **Calendar → tradingDays → option** specifies the type of calendar applied to the series, with possible settings being **None** (no calendar variables), **Default** (the standard JDemetra+ calendar without national holidays), **Stock** (day-of-week effects for end-of-period data such as inventories), **Holidays** (predefined trading-day variables adjusted for country-specific holidays), and **UserDefined** (trading-day variables from a custom list).
- **Calendar → tradingDays → td** specifies the type of regression effect assigned to the predefined calendar variables. It is available only when **Calendar → tradingDays → option** is set to **Default** or **Holidays**, with possible values being **None** (no calendar variables), **tradingDays** (six day-of-the-week regressors), and **WorkingDays** (a working versus non-working day contrast).
- **Calendar → tradingDays → Leap year** controls whether and how the leap-year effect is incorporated into the model, with available values being **None** (no leap-year adjustment), **LeapYear** (a contrast variable for leap years), and **LengthOfPeriod** (a regression variable reflecting the length of the month or quarter).

- **Calendar → tradingDays → autoAdjust** enables automatic correction for the leap-year effect, available only when in the **Transformation** section **function** is set to **Auto**.
- **Calendar → tradingDays → test** performs a pre-test of the trading-day regression variables using the AICC statistic, with options including **Add** (variables excluded initially but added if significant), **None** (no pre-test, variables always included), and **Remove** (variables included initially but dropped if not significant). While this option does not affect calendar tests, it may indirectly influence other procedures such as the log/level test.
- **Calendar → easter → IsEnabled** activates the option to account for the Easter effect in the RegARIMA model, meaning that a correction is applied when selected, with the actual inclusion depending on the **Pretest** settings.
- **Calendar → easter → Julian** enables inclusion of the Easter effect based on the Julian calendar, converted to the corresponding Gregorian date, with the checkbox unmarked by default.
- **Calendar → easter → Pre-test** evaluates the significance of the Easter regression variable using the AICC statistic, with options **Add** (initially excluded but added if significant), **None** (no pre-test, effect always included), and **Remove** (initially included but dropped if not significant).
- **Calendar → easter → Easter duration** defines the length of the Easter effect, which can be set between 1 and 20 days, is available only when **Calendar → easter → Pre-test** is set to **None** or **Remove**, and has a default value of 8.

The **Pre-specified outliers** section allows the user to include outliers that are known in advance, rather than relying solely on automatic detection. Users can define the type of outlier, its timing, and its magnitude so that it is explicitly incorporated into the model. Typical outlier types include Additive Outliers (AO), Level Shifts (LS), Transitory Changes (TC), and Seasonal Outliers (SO) (Figure 5.1.7).

Figure 5.1.7: Types of Outliers

Source: International Monetary Fund. 2017. *Quarterly National Accounts Manual*. Washington, DC: IMF.

Each of these outlier types represents a distinct kind of irregular movement in a time series. AO corresponds to one-off shocks that affect only a single observation, such as data entry errors or extraordinary events. LS indicates a permanent change in the series level, reflecting structural breaks like reclassification of data or major policy reforms. TC describes short-lived deviations followed by a gradual return to the underlying trend, often caused by temporary shocks such as extreme weather or strikes. SO affects only specific periods of the year and may result from unusual seasonal patterns, for instance, changes in holiday timing or irregular sales promotions.

Proper identification and treatment of these outliers are important, as they can distort the estimation of seasonal and trend components, leading to misleading interpretations of economic movements. When incorporating outliers into a seasonal adjustment model, it is important to ensure that each identified outlier has a clear economic rationale and is statistically significant within the model. Outliers should not be retained solely because they improve statistical fit; instead, they must correspond to identifiable events such as policy changes, strikes, disaster triggered by natural hazard, or data collection issues.[11] Once confirmed, these outliers should be fixed under the **Regression → Pre-specified outliers** section in JDemetra+, ensuring they are consistently applied in future model runs. After specifying the type, timing, and magnitude of each outlier, the model should be saved to preserve these settings for reproducibility and future reference.

Intervention variables are used to model the impact of events that produce structural changes in a time series. These may represent temporary shocks, permanent shifts, or gradual transitions. JDemetra+ allows the user to define interventions at specific dates and choose their functional form (e.g., pulse, step, or ramp). Such variables are especially valuable for capturing effects of extraordinary events, such as the introduction of new legislation or a sudden economic crisis that cannot be explained by regular seasonal or calendar patterns.

Ramp effects provide a way to model gradual changes in a series over a defined time interval. Unlike a one-time intervention, a ramp variable assumes a linear increase or decrease between a start and end date, after which the effect stabilizes. This specification is often applied in situations where changes are phased in over time, such as the rollout of new regulations, gradual adoption of technology, or long-lasting policy reforms. The bottom-right panel in Figure 5.1.7 illustrates a ramp example, showing a smooth upward adjustment over several periods before reaching a new stable level.

The **User-defined variables** option offers full flexibility by allowing users to introduce their own regression variables into the model. These variables must be created beforehand and can represent any deterministic factor relevant to the series, such as institution-specific adjustments, country-specific holiday calendars not included in the default setup, or sector-specific dummies. By incorporating user-defined variables, analysts can tailor the RegARIMA model to the precise needs of their data and institutional context.

Outliers

The next part of the specification is the **Outliers** section (Figure 5.1.8), which allows automatic detection of irregular observations in the series.[12] Depending on the settings, the program can identify additive outliers, level shifts, temporary changes, seasonal outliers, or any combination of these, based on the chosen model. When the option **Is enabled** is selected, which is the case in all default specifications, outlier detection and correction are carried out during the seasonal adjustment process.

11 D. Ladiray and B. Quenneville. 2001. *Seasonal Adjustment with the X-11 Method.* New York: Springer.

12 D. F. Findley and B. C. Monsell. 2009. Model Selection and Outlier Detection in Seasonal Adjustment. *Journal of Official Statistics.* 25 (1). pp. 35-70.

The **Detection span → Type** parameter specifies the portion of the time series to be examined for outliers, while the **Use default critical value** option sets the detection threshold. This threshold is the main fine-tuning control: raising the value lowers the number of detected outliers, whereas lowering it increases sensitivity. It is generally advisable to begin with the default setting, though adjustments may be useful for problematic series. Additional checkboxes allow the user to specify which outlier categories (AO, LS, TC, SO) should be included in the search. Two further options, **TC rate** and **Method**, are also available, but since they are not relevant for standard seasonal adjustment procedures, they are usually left at their default settings.

Figure 5.1.8: Outliers

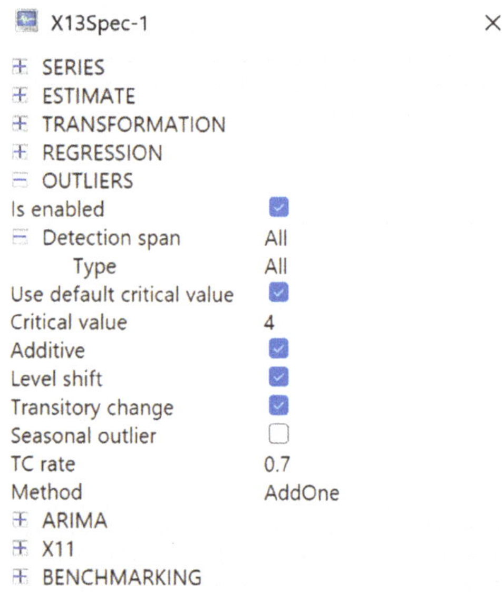

TC = transitory changes.

Source: Screenshot by the Asian Development Bank based on JDemetra+ 2.2.2.

The default critical value applied in outlier detection depends on the length of the time series, as established in the X-12-ARIMA Reference Manual (Version 0.3). This value determines the threshold beyond which an observation is classified as a potential outlier. In shorter series, lower critical values are used to maintain detection sensitivity, while in longer series higher values are applied to prevent excessive false positives. As shown in the X-12-ARIMA documentation, the critical value increases gradually with the number of observations tested—from 1.96 for a single observation to around 4.07 for a 360-observation series. For example, a series with 72 observations has a default critical value of 3.73, while one with 240 observations uses 3.99. This scaling ensures that the probability of incorrectly identifying an outlier remains approximately constant regardless of sample size. Although JDemetra+ automatically assigns these defaults, users may adjust the critical value manually when the automatic detection either misses obvious irregularities or produces too many spurious outliers.

ARIMA

The **ARIMA** panel governs the identification of the ARIMA part of the RegARIMA model (Figure 5.1.9). The model structure can be specified manually or determined through automatic identification. When the **Automatic** checkbox is selected, JDemetra+ determines the ARIMA structure, limiting non-seasonal polynomials to order 3 and seasonal polynomials to order 1. The following options are available for automatic model identification:

- **Accept Default:** Checks whether the default airline model ARIMA(0,1,1)(0,1,1) is adequate by applying the Ljung-Box Q-statistic to its residuals.[13] If acceptable, this model is retained without further search.
- **Cancellation limit:** Sets the tolerance for treating autoregressive (AR) and moving average (MA) roots as equal, mainly used in determining differencing orders. The default is 0.1.

[13] G. M. Ljung and G. E. P. Box. 1980. Analysis of Variance with Autocorrelated Observations. *Scandinavian Journal of Statistics.* 7 (3). pp. 172–180.

- **Initial UR (Diff.):** Threshold for the first unit root test in differencing order selection. If, in the preliminary model (2,0,0)(1,0,0) with mean, a root exceeds the limit in modulus, it is set to unity. Default = 1.0417.
- **Final UR (Diff.):** Threshold for the second unit root test. Applied to the (1,0,1)(1,0,1) with mean model, it checks whether a root exceeds the limit; if so, differencing is increased unless cancellation is possible.[14] Default = 0.88.
- **Mixed:** Determines whether models with both seasonal and non-seasonal AR and MA terms are permitted. If disabled, AR and MA terms cannot appear simultaneously in both seasonal and non-seasonal parts.
- **Balanced:** Gives preference to balanced models, where the combined AR and differencing order equals the combined MA order. By default, this option is off; when enabled, it mirrors TRAMO's preference.

Figure 5.1.9: ARIMA

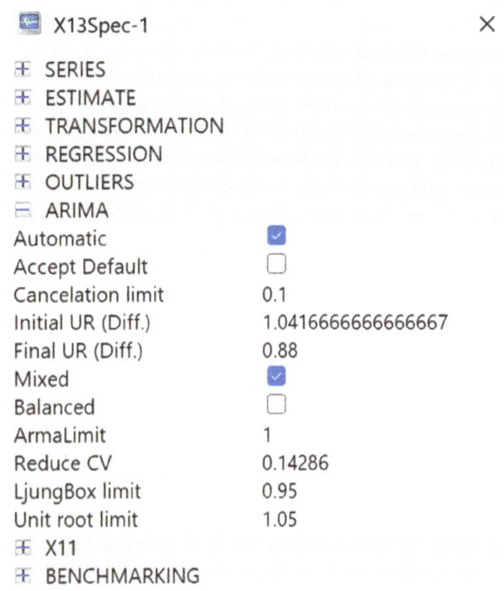

UR = unit root.

Source: Screenshot by the Asian Development Bank based on JDemetra+ 2.2.2.

- **ArmaLimit:** t-statistic threshold for ARMA coefficients. If the highest-order coefficient has a t-value below the limit, the order is reduced. This rule also applies to the constant term. Default = 1.
- **Reduce CV:** Proportion by which the outlier critical value is lowered if the Ljung-Box Q-statistic fails. Active only with automatic outlier detection. Default = 0.14268.
- **LjungBox limit:** Acceptance level for the Ljung-Box Q-statistic (lag 24 for monthly data, 16 for quarterly). If exceeded, the model is rejected, the outlier critical value is reduced, and the procedure is repeated. Default = 0.95.
- **Unit root limit:** Final unit root test threshold. If the absolute value of an AR root is below this limit, a unit root is assumed, AR order reduced, and differencing order increased.[15] Default = 1.05.

Benchmarking

The **Benchmarking** panel allows users to reconcile seasonally adjusted results with a benchmark series to ensure consistency across different frequencies (Figure 5.1.10). **Benchmarking** can be activated by selecting the **Is enabled** checkbox. Once enabled, the user specifies the **Target**, typically the calendar-adjusted series against which the results will be aligned. Additional options include **Use forecasts**, which extends the adjustment by incorporating forecasted values, and the parameters **Rho** and **Lambda**, which control the strength and smoothness of the adjustment process. Both parameters are set to 1 by default, but can be modified if greater flexibility in the benchmarking process is required.

14 W. R. Bell and S. C. Hillmer. 1984. Issues Involved with the Seasonal Adjustment of Economic Time Series. *Journal of Business & Economic Statistics.* 2 (4). pp. 291–320.

15 G. E. P. Box and G. M. Jenkins. 1976. *Time Series Analysis: Forecasting and Control.* San Francisco: Holden-Day.

By default, benchmarking is turned off, as it is not always necessary, but it becomes essential in institutional contexts where monthly or quarterly results must be made coherent with officially published annual totals.[16]

Figure 5.1.10: Benchmarking

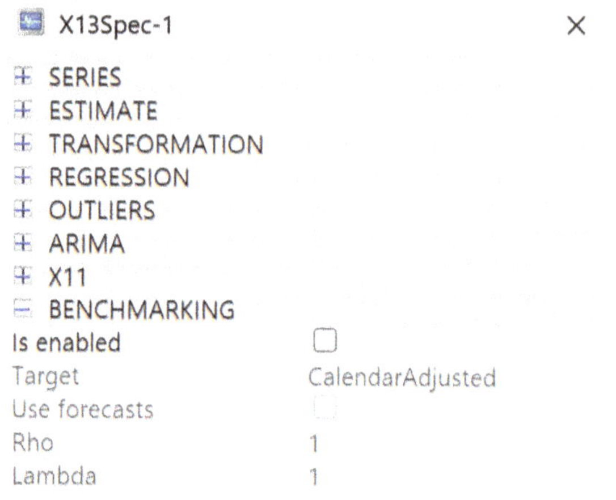

Source: Screenshot by the Asian Development Bank based on JDemetra+ 2.2.2.

5.2 Single Processing

Single processing refers to applying seasonal adjustment to a single time series at a time, using a chosen specification. This approach is particularly useful when working closely with one dataset, as it allows the user to fine-tune settings, test alternative specifications, and immediately view the results.

Once the desired specification has been selected, a single-process seasonal adjustment can be launched in two different ways. If a default specification (for example, RSA5c for X-13) is used, the procedure is started through the main menu by choosing **Statistical Methods → Seasonal Adjustment → Single Analysis → X13**. This action creates a new blank document in **Workspace → Seasonal Adjustment → Documents**, into which the target time series can be dragged and dropped. Another way to launch the seasonal adjustment, when working with a pre-modified specification, is to right-click the specification of interest under **Workspace → Seasonal Adjustment → Specifications → X13** and choose **Create Document** (Figure 5.2.1).

Figure 5.2.1: Single-Process Document Creation

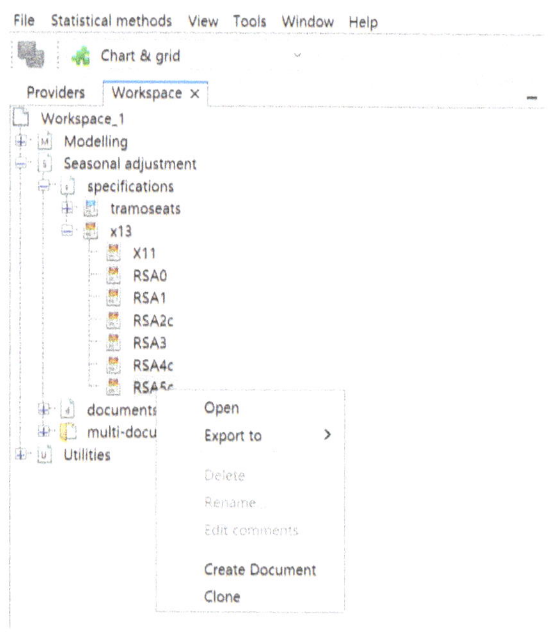

Source: Screenshot by the Asian Development Bank based on JDemetra+ 2.2.2.

The exact sequence of actions involved in a single-process seasonal adjustment is illustrated in the figures below. In practice, the workflow begins when the dataset is called by **Providers** and the relevant time series is identified. Next, a user-defined calendar is created and national holidays are entered into the calendar module.

16 B. Quenneville, P. A. Cholette, E. B. Dagum, and B. Rioux. 2004. Adjustment of Seasonally Adjusted Series to Annual Totals. *Proceedings of the American Statistical Association, Business and Economic Statistics Section*, 1–8; P. A. Cholette and E. B. Dagum. 1994. Benchmarking Time Series with Autocorrelated Survey Errors. *International Statistical Review*. 62 (3). pp. 365–377; E. B. Dagum and P. A. Cholette. 2006. *Benchmarking, Temporal Distribution, and Reconciliation Methods for Time Series*. New York: Springer.

A default specification is then cloned to serve as the basis for a customized specification, after which the cloned version is modified as required. Once these preparations are complete, a new seasonal adjustment document is created, producing a blank workspace ready for the time series to be dragged and dropped. After all these steps have been carried out, the results of the single-process seasonal adjustment can be viewed and further analyzed.

Step 1: Selecting time series

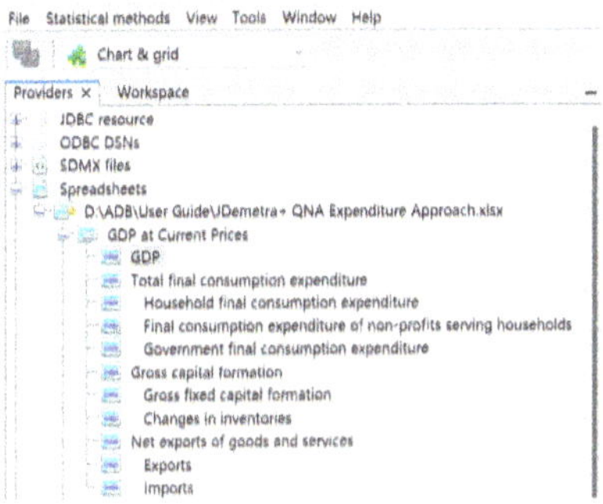

Step 2: Adding national calendar

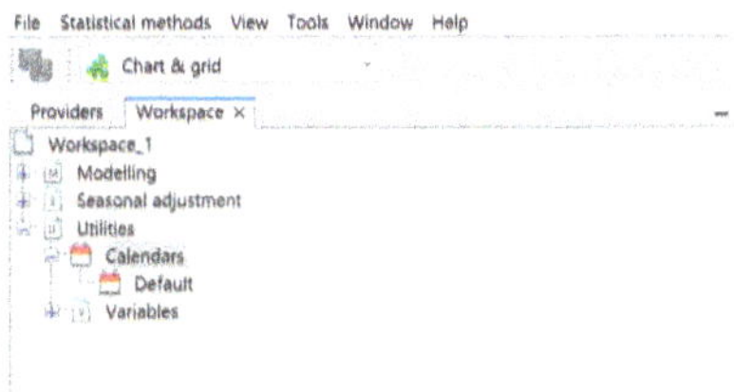

Step 3: Adding national holidays

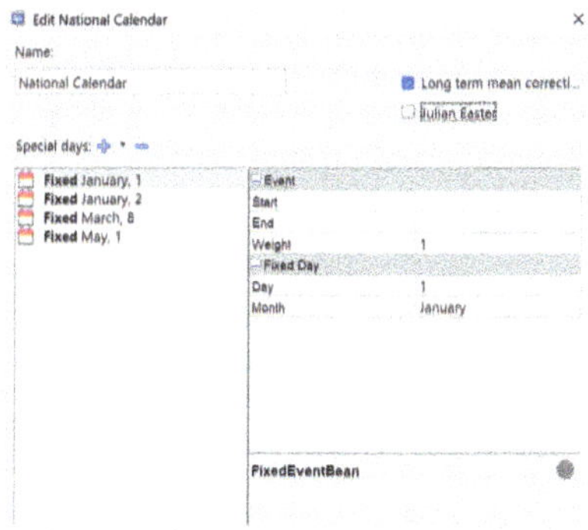

Step 4: Cloning a specification

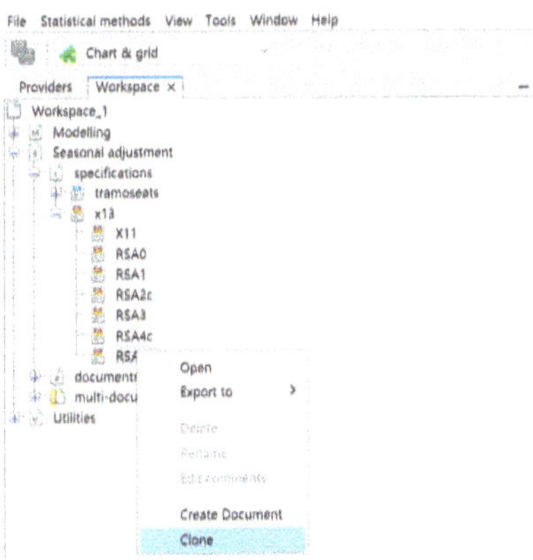

Step 5: Setting up the specification

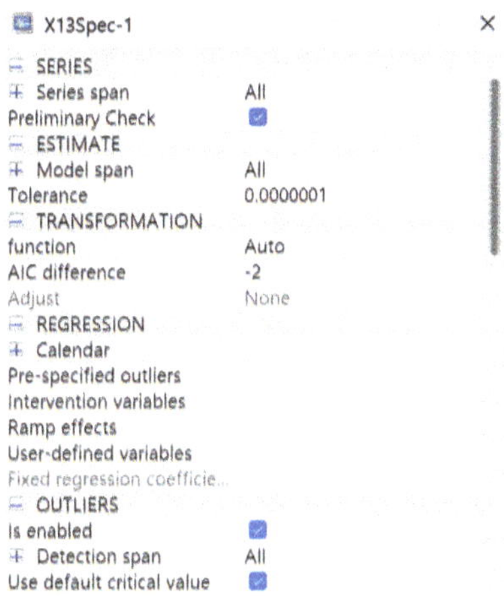

Step 6: Creating a blank document

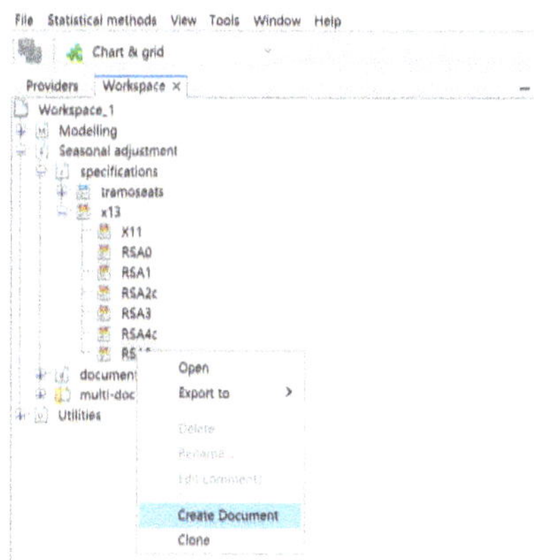

Step 7: Getting results

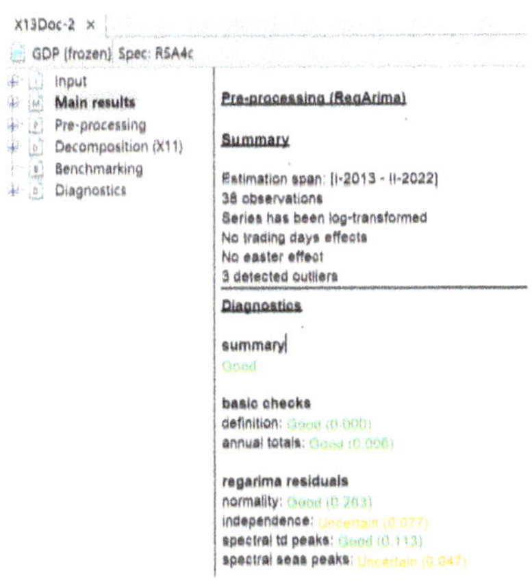

5.3 Multi-Processing

The multi-processing function in JDemetra+ makes it possible to seasonally adjust multiple time series at once, with the flexibility of applying different specifications and methods to each individual series. A new multi-processing document can be created by selecting **Statistical Methods → Seasonal Adjustment → Multi-processing → New** from the main menu (Figure 5.3.1, left). This action opens a blank document, where the series to be adjusted can be added by dragging and dropping, similar to the procedure used in single processing (Figure 5.3.1, right).

Figure 5.3.1: Multi-Process Document Creation

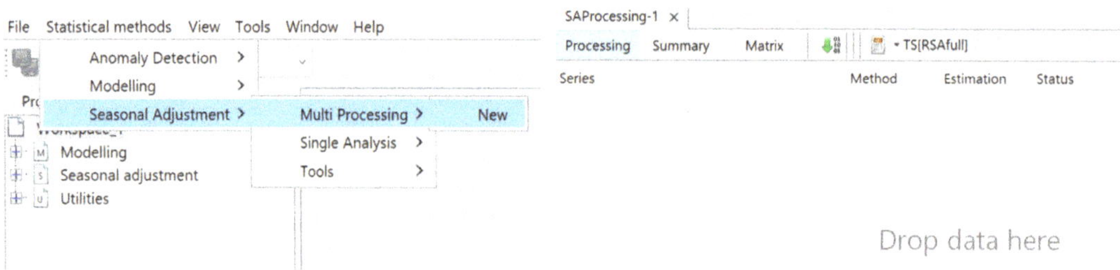

Source: Screenshot by the Asian Development Bank based on JDemetra+ 2.2.2.

Once the time series have been added to the multi-processing document, the next step is to assign a specification to each of them. This can be a standard template, such as **RSA5c** for X13, or a user-defined option tailored to the characteristics of the series. For example, in the case of gross domestic product, a customized specification named **User** has been created and applied in the figures below.

Step 1: Creating a blank document

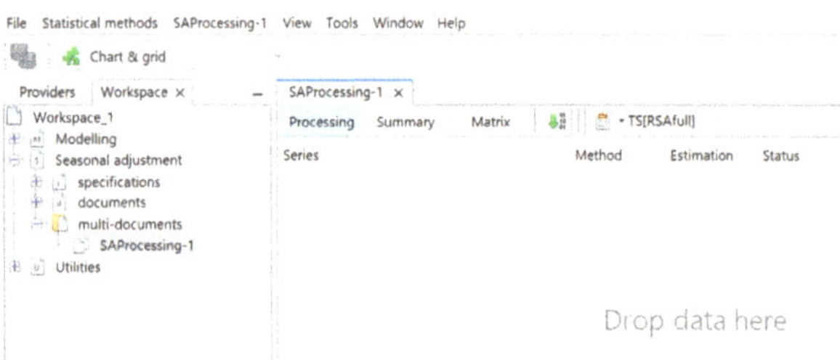

Step 2: Dragging and dropping data

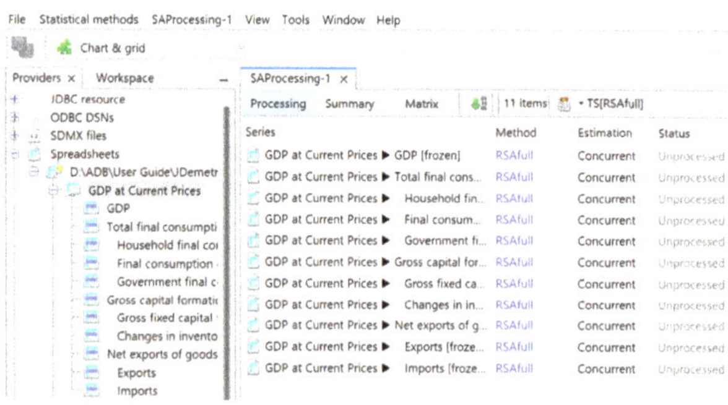

Step 3: Choosing the desired specification

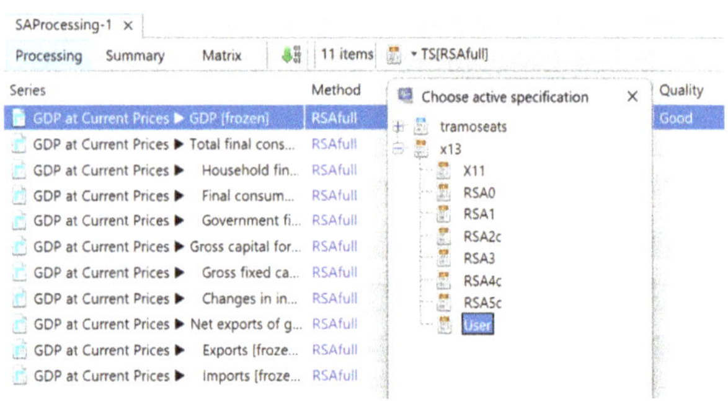

Step 4: Saving the desired specification

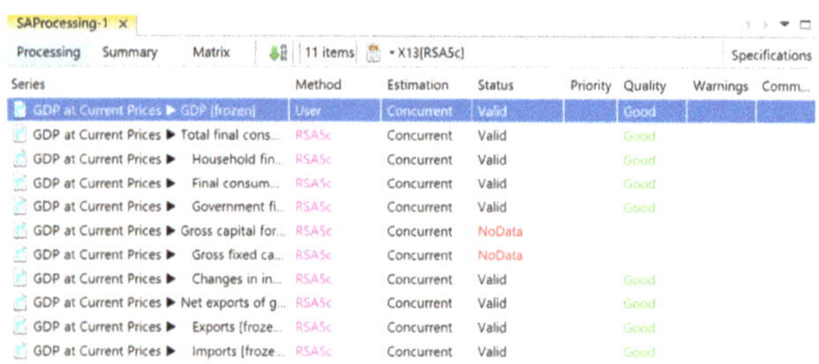

Assigning specifications in this way allows the analyst to mix and match default and customized approaches within the same framework, ensuring that each series is adjusted according to its unique features. After the specifications have been defined, the adjustment is executed, and the results are displayed in a structure identical to that of single processing.

One of the key advantages of multi-processing is the ability to duplicate a series and process it under different specifications. In practice, this means that gross domestic product can be adjusted once using the default **RSA5c** template and again using the **User** specification. The results of the two adjustments can then be reviewed side by side, providing insights into how methodological choices affect final outcomes. This functionality is particularly useful for testing robustness, comparing the performance of competing models, or meeting institutional requirements that call for evaluation under multiple scenarios. In this way, multi-processing not only increases efficiency when handling large sets of data but also supports more rigorous and transparent decision-making in seasonal adjustment.

ANALYSIS OF THE RESULTS

6

6.1 Main Results

Once the seasonal adjustment procedure has been carried out, JDemetra+ provides a concise set of outputs summarizing both the model applied and the quality of the results. These outputs help the analyst quickly verify whether the decomposition is satisfactory and whether the assumptions underlying the adjustment are supported by diagnostic tests. In practice, the “Main results” panel in JDemetra+ combines essential information on the adjustment span, presence of calendar effects, number of outliers, and a series of diagnostic indicators covering residual properties, stability checks, and statistical tests for seasonality. The goal is to provide a transparent overview that can be easily interpreted without delving into the full set of detailed outputs.

Figure 6.1.1 shows the main results table for a hypothetical series adjusted with the RSA5c specification described in Section 5. The estimation period runs from the first quarter of 2013 to the second quarter of 2022, giving a total of 38 observations. In this example, no trading-day or Easter effects were detected, while three outliers were identified during pre-adjustment. The overall model summary indicates that the adjustment is acceptable. The basic checks confirm that the decomposition is internally consistent and that the seasonally adjusted series matches the annual totals without discrepancy.

Figure 6.1.1: Main Results Window

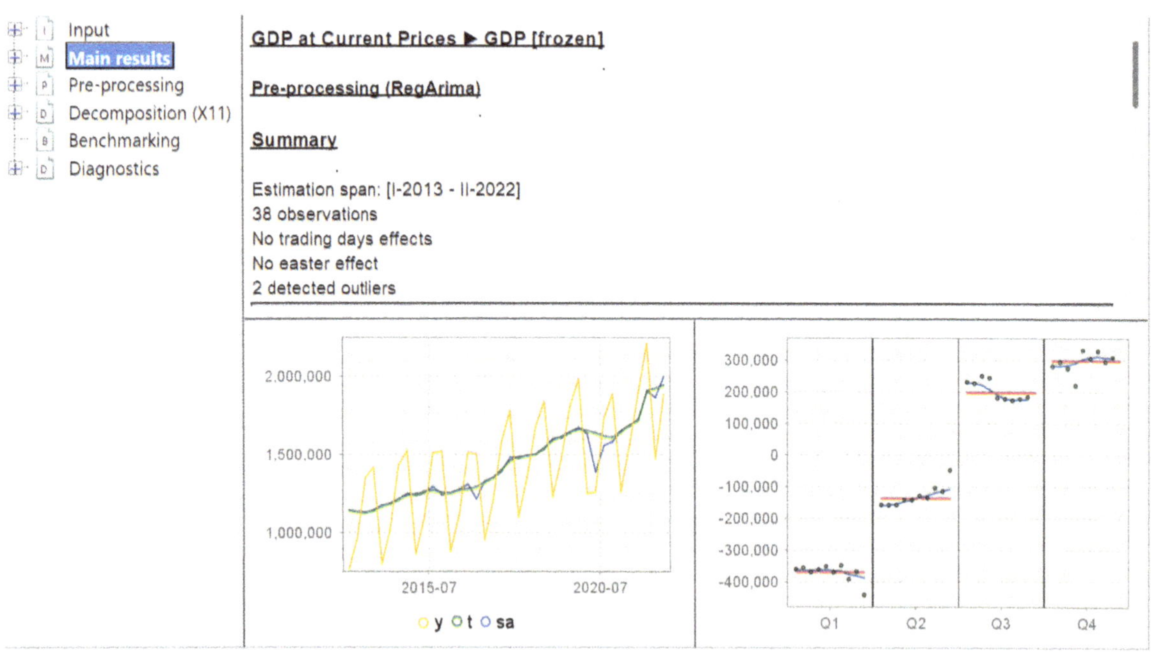

GDP = gross domestic product.
Source: Screenshot by the Asian Development Bank based on JDemetra+ 2.2.2.

The residual diagnostics provide further assurance about model adequacy. The normality test (0.730) checks whether the residuals are approximately normally distributed, while the independence test (0.561) verifies the absence of autocorrelation; both values indicate that the residuals behave as expected.
The spectral tests show no significant trading-day peaks (0.356) or seasonal peaks (0.331), suggesting that calendar and seasonal effects have been adequately removed.

The outlier section shows three irregular points in the data, with the outlier statistic (0.079) highlighting that this aspect of the model is somewhat weaker than others. However, the M-statistics (Q = 0.230, Q-M2 = 0.260) confirm the stability of the adjustment, as these indicators measure how well the decomposition captures underlying structure without leaving residual seasonality.

The residual seasonality tests also support the robustness of the adjustment. The QS test on the seasonally adjusted series (1.000) and the irregular component (1.000) suggest no residual seasonal pattern, while the F-tests on seasonal dummies (0.852 for the adjusted series and 0.605 for the irregular component) confirm that seasonality has been effectively removed. Finally, the residual trading-day tests (0.223 for the adjusted series and 0.102 for the irregular component) show that no trading-day effect remains in the data after adjustment.

6.2 Visual Check

A quick way to evaluate the outcome of a seasonal adjustment in JDemetra+ is by examining the graphical outputs that accompany the main results. The program automatically generates a set of charts and tables that allow users to interpret the decomposition visually. As shown in Figure 6.1.1, the panel on the left combines three series: the raw data, the estimated trend-cycle, and the seasonally adjusted values.
On the right, the software presents the **Seasonal-Irregular (S-I) ratio**, which shows how much variation is left unexplained after removing the trend. These charts are interactive and can be explored in more detail. Double-clicking on a particular line in the left-hand panel isolates that component and opens it in a separate window, while double-clicking within the S-I ratio panel zooms in on the selected quarter or month for a more detailed view. This functionality allows users to examine either long-term trends or individual time periods without cluttering the main display.

Additional visualizations are available under the **Charts** section of the **Main Results**. The option **“Sa, trend”** displays the original series alongside the trend-cycle and the seasonally adjusted values, as shown in the left panel of Figure 6.2.1. This chart makes it possible to see how much of the short-term variation is removed through adjustment and how the underlying trend is revealed once seasonal effects are filtered out.
By selecting the option labelled **“Cal., sea., irr.”**, the software produces a second visualization that breaks the series into calendar effects (yellow), seasonal fluctuations (green), and irregular movements (blue). The right panel of Figure 6.2.1 provides an example of this decomposition, which helps analysts identify how much of the observed variation is due to systematic seasonal or calendar patterns and how much is explained by irregular shocks.

Figure 6.2.1: Charts Node

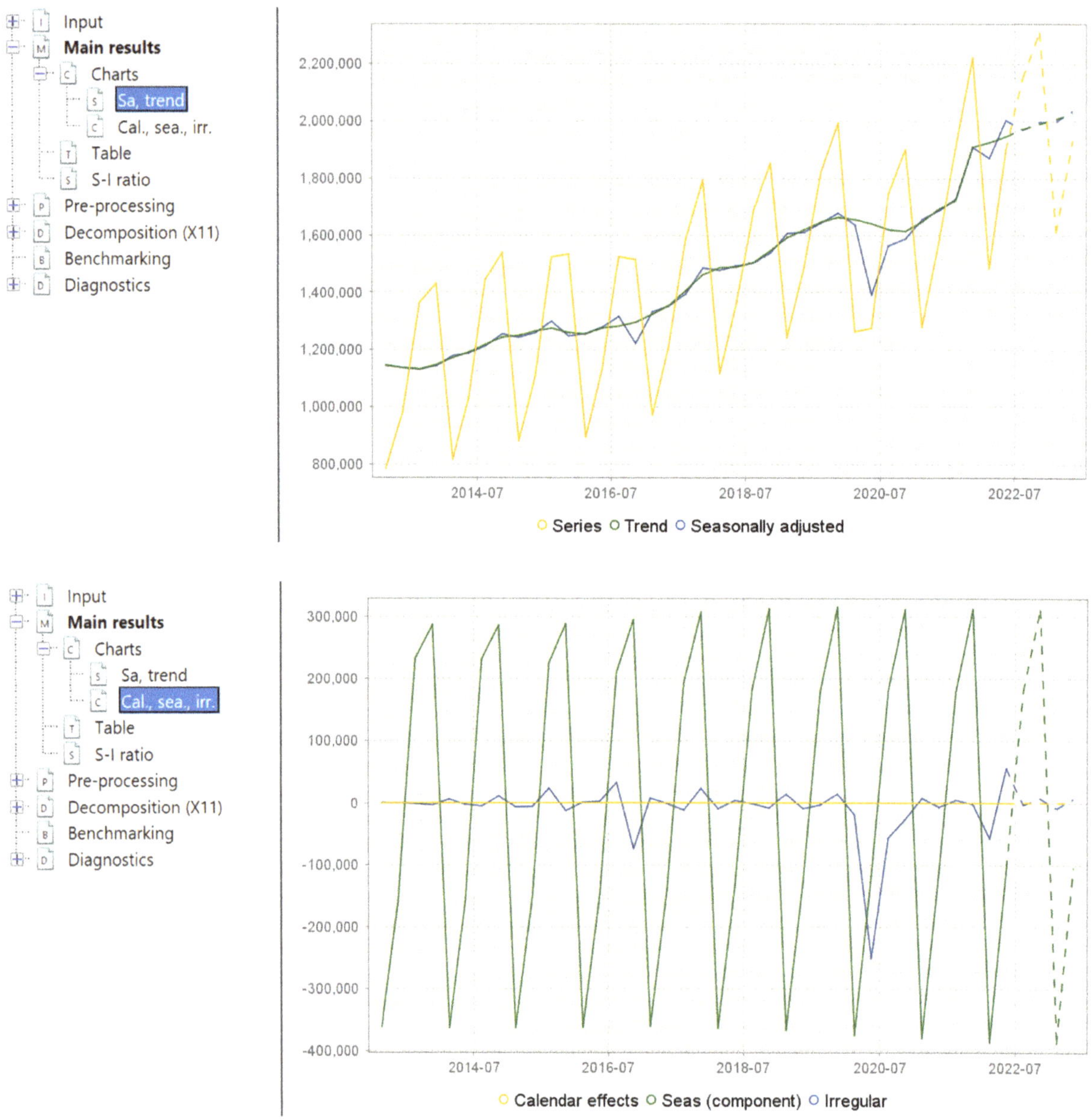

Source: Screenshot by the Asian Development Bank based on JDemetra+ 2.2.2.

Another important visualization in JDemetra+ is the **S-I ratio**, accessible under the **Charts** node. Figure 6.2.2 illustrates this diagnostic, which plots the seasonal component relative to the irregular component for each quarter (or month, if monthly data). In the chart, the black dots represent the observed S-I ratios for each period, the blue line traces the smoothed seasonal pattern, and the red line shows the average level within each quarter. The S-I ratio is particularly valuable because it highlights whether the estimated seasonal component is stable or dominated by irregular fluctuations. A relatively smooth blue curve that remains close to the red reference line suggests that the seasonal adjustment is reliable, while large swings or instability point to residual seasonality or possible outliers.

Figure 6.2.2: Seasonal-Irregular Ratio

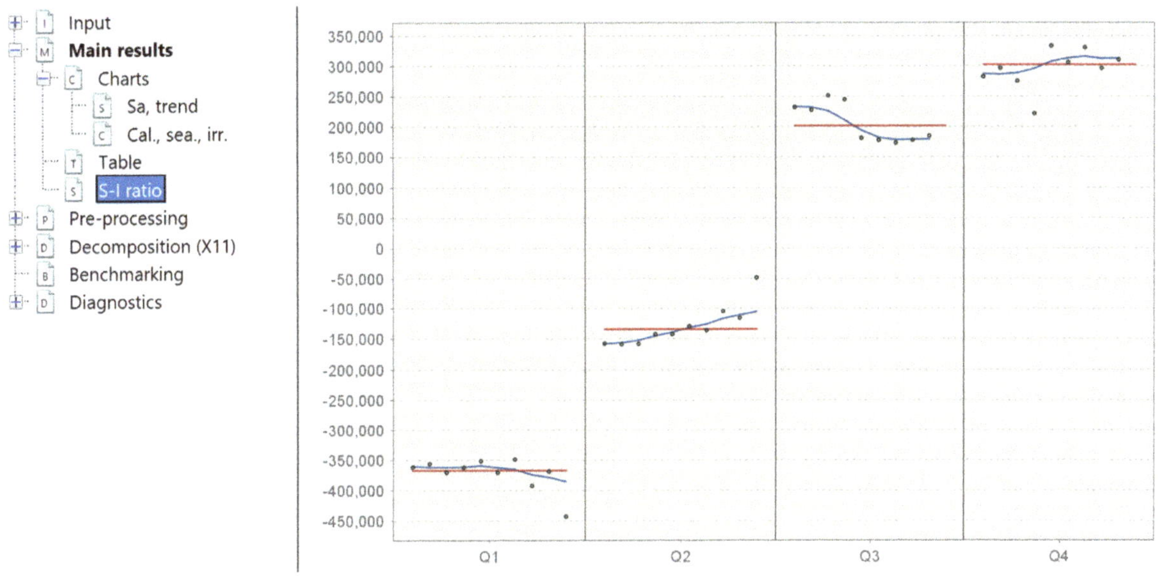

Source: Screenshot by the Asian Development Bank based on JDemetra+ 2.2.2.

6.3 Pre-Processing of Series

The pre-processing stage plays a central role in preparing a time series for seasonal adjustment. It consists of a summary of the ARIMA model estimation, which includes key details such as the estimation span used, data transformations applied, the specific ARIMA model selected, and the parameters associated with trading days, user-defined regressors, Easter effects, and outliers. This section also contains several sub-nodes—**Forecasts**, **Regressors**, **ARIMA**, **Pre-adjustment series**, **Residuals**, and **Likelihood**—each of which provides important information for assessing the adequacy and robustness of the pre-processing model.

Forecasts

A key step in validating the quality of a seasonal adjustment model is the analysis of out-of-sample forecasts.[17] In X-13, the identified ARIMA model is used to generate point forecasts, forecast standard errors, and prediction intervals. The prediction interval on the transformed scale takes the form of the point forecast plus or minus a multiple of the forecast standard error. The multiplier is drawn from the normal distribution according to the desired coverage probability. By default, X-13 displays a 95% prediction interval, which corresponds to a multiplier of 1.96. The forecasting procedure assumes that no new outliers emerge during the forecast horizon. The resulting forecasts are presented in the Forecast node, where the observed series is shown together with projections for the next year and the associated prediction interval, as illustrated in Figure 6.3.1.

[17] D. Peña, G. C. Tiao, and R. S. Tsay. 2001. *A Course in Time Series Analysis.* New York: John Wiley & Sons.

Figure 6.3.1: Forecast Panel

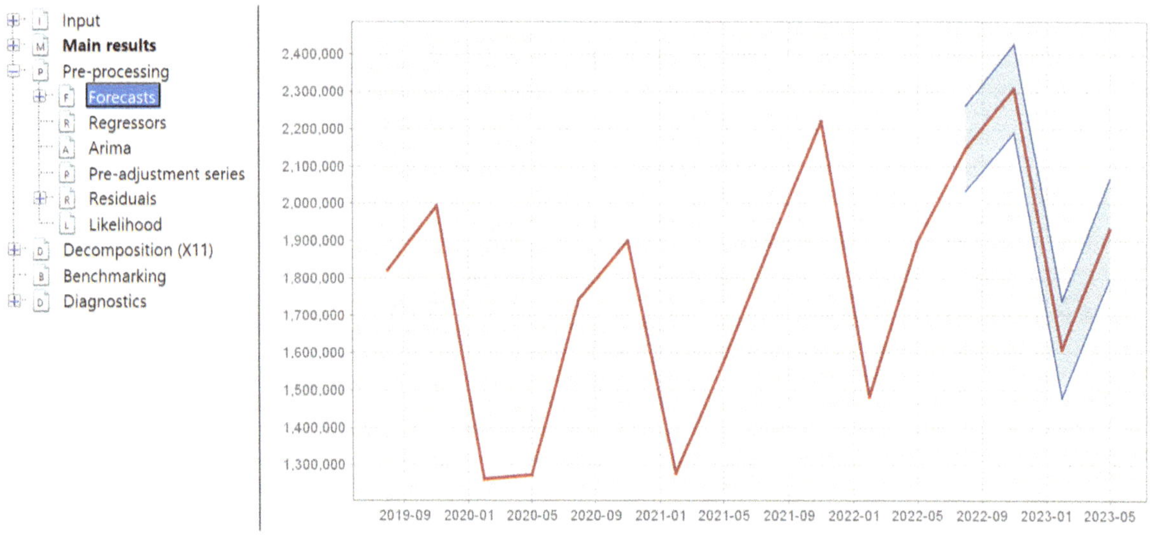

Source: Screenshot by the Asian Development Bank based on JDemetra+ 2.2.2.

The **Forecast** node is divided into two subsections: **Table** and **Out-of-sample test**. The table section presents the forecasts as a table, where users can view the numerical values directly, reformat them for easier inspection, or export them to other applications. This functionality allows the forecasts to be integrated into further analysis or reporting, while also offering flexibility in terms of visualization and layout. The second subsection focuses on out-of-sample testing, in which the historical series is split into a fit period, used to estimate the model, and a test period, withheld for validation. The model is then applied to generate one-step-ahead forecasts for the test period, and these are compared with the actual observations. The evaluation examines whether forecast errors are unbiased and whether their variance is consistent with the variance of the in-sample residuals. When the two measures are broadly aligned, the model is considered adequate, whereas large discrepancies point to potential misspecification.

Regressors

The **Regressors** section provides a complete overview of all deterministic variables incorporated into the pre-processing model. These typically include trading-day adjustments, the leap year effect, calendar-related holidays such as Easter, and various intervention variables designed to capture structural breaks or exceptional events. Outliers, ramps, and any user-defined regressors specified by the analyst are also displayed in this window, as shown in Figure 6.3.2. For practical purposes, the **Regressors** window allows users to copy series directly for use in other applications. A single regressor can be selected and copied by clicking its header and using the standard keyboard shortcut, while the entire table can be exported by selecting the top-left empty cell. This functionality is particularly useful when analysts wish to further examine regressor values in statistical software or spreadsheet environments.

Figure 6.3.2: Regressors Window

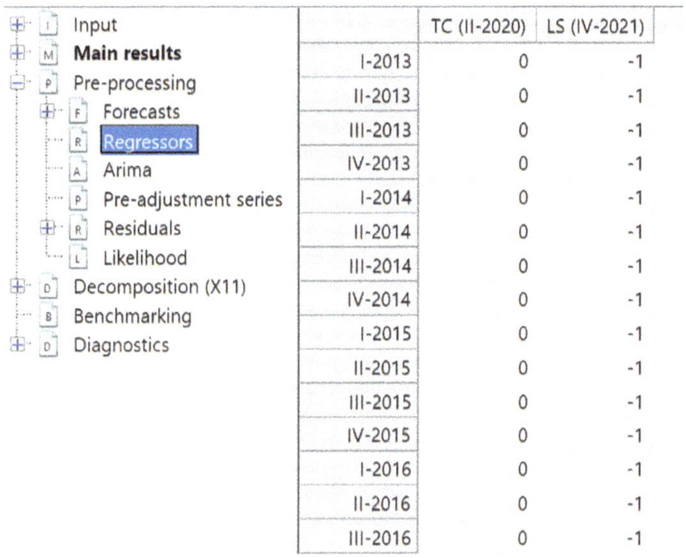

	TC (II-2020)	LS (IV-2021)
I-2013	0	-1
II-2013	0	-1
III-2013	0	-1
IV-2013	0	-1
I-2014	0	-1
II-2014	0	-1
III-2014	0	-1
IV-2014	0	-1
I-2015	0	-1
II-2015	0	-1
III-2015	0	-1
IV-2015	0	-1
I-2016	0	-1
II-2016	0	-1
III-2016	0	-1

Source: Screenshot by the Asian Development Bank based on JDemetra+ 2.2.2.

ARIMA

The **ARIMA** section (Figure 6.3.3) presents detailed information on the statistical properties of the model selected during pre-processing. It combines theoretical and empirical aspects of the specification, distinguishing between stationary and non-stationary representations. In the lower part of the panel, the chosen ARIMA model is displayed in symbolic form, with parameters capturing the regular autoregressive (AR), differencing (I), and moving average (MA) terms, together with their seasonal counterparts. The estimated coefficients for the AR and MA terms are expressed using the backshift operator B, while the roots of the characteristic equation are also reported for regular AR components, along with their argument and modulus.

Pre-Adjusted Series

Figure 6.3.4 displays a range of series estimated by X-13ARIMA-SEATS, the exact contents of which depend on the regressors included in the model. These series represent different stages of adjustment applied to the raw data, helping to isolate the influence of calendar effects, outliers, and other deterministic components before seasonal adjustment is carried out.

Among the outputs, users will often find versions of the series that have been interpolated for missing values, transformed, or linearized, or corrected for calendar-related influences such as trading days, leap years, or moving holidays like Easter. The model also decomposes the deterministic effects more explicitly, showing the contribution of components such as ramps, interventions, or level shifts, as well as the combined impact of calendar variables. In addition, the table reports the effects of outliers on different parts of the decomposition—trend, seasonal, and irregular components—allowing users to assess how much of the variation in the series is attributable to unusual events. Similarly, regression effects are displayed for various components depending on how user-defined variables were specified in the model.

Figure 6.3.3: ARIMA Window

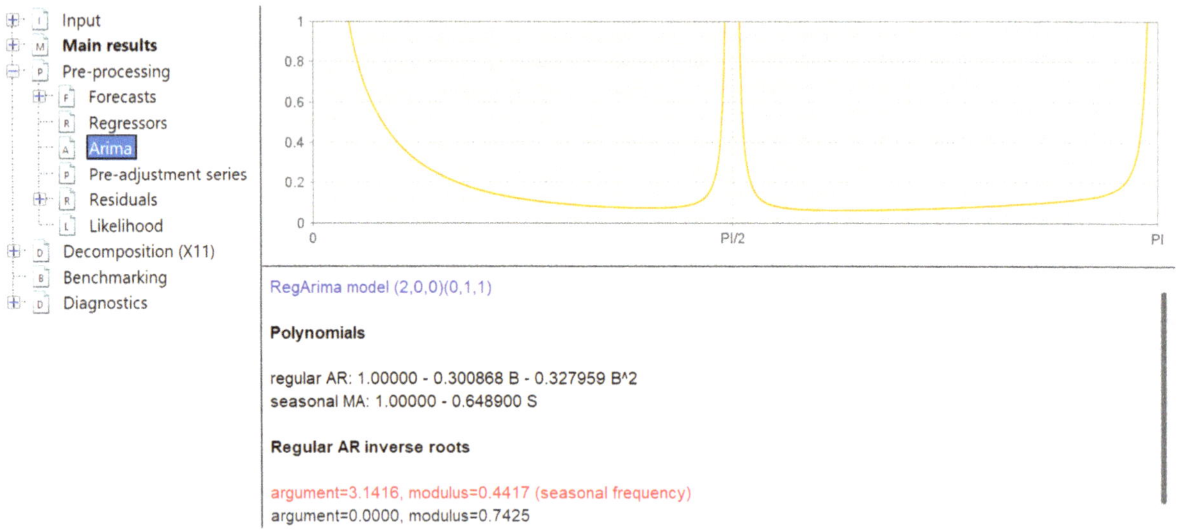

Source: Screenshot by the Asian Development Bank based on JDemetra+ 2.2.2.

Figure 6.3.4: Pre-Adjusted Series

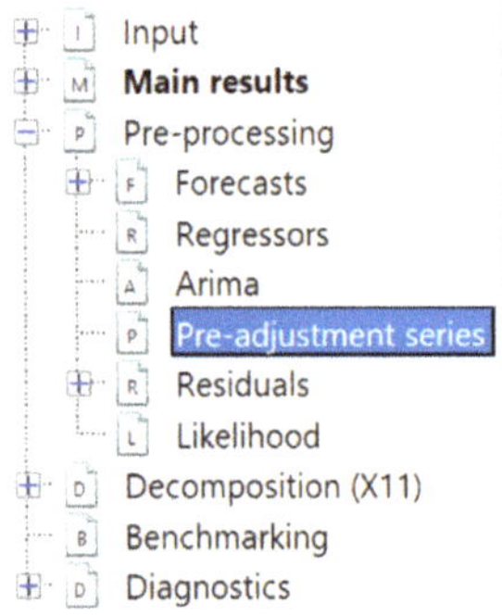

	yc	y_lin	ycal	det	cal	omhe	tde
I-2013	784,166.2	951,305.723	784,166.2	-167,139.5...	0	0	0
II-2013	978,205.1	1,145,344....	978,205.1	-167,139.5...	0	0	0
III-2013	1,363,072.3	1,530,211....	1,363,072.3	-167,139.5...	0	0	0
IV-2013	1,430,194.6	1,597,334....	1,430,194.6	-167,139.5...	0	0	0
I-2014	814,938.4	982,077.923	814,938.4	-167,139.5...	0	0	0
II-2014	1,030,592.9	1,197,732....	1,030,592.9	-167,139.5...	0	0	0
III-2014	1,443,501.1	1,610,640....	1,443,501.1	-167,139.5...	0	0	0
IV-2014	1,539,593.9	1,706,733....	1,539,593.9	-167,139.5...	0	0	0
I-2015	879,565.7	1,046,705....	879,565.7	-167,139.5...	0	0	0
II-2015	1,105,248	1,272,387....	1,105,248	-167,139.5...	0	0	0
III-2015	1,523,902.7	1,691,042....	1,523,902.7	-167,139.5...	0	0	0
IV-2015	1,534,916.8	1,702,056....	1,534,916.8	-167,139.5...	0	0	0
I-2016	893,634.2	1,060,773....	893,634.2	-167,139.5...	0	0	0
II-2016	1,132,260.6	1,299,400....	1,132,260.6	-167,139.5...	0	0	0
III-2016	1,525,628.4	1,692,767....	1,525,628.4	-167,139.5...	0	0	0
IV-2016	1,515,770.3	1,682,909....	1,515,770.3	-167,139.5...	0	0	0

Source: Screenshot by the Asian Development Bank based on JDemetra+ 2.2.2.

Residuals

The **Residual** section is intended to verify whether the errors from the ARIMA model behave as pure random noise. This means that residuals should be normally distributed, independent, random in their sequence, and free of any remaining linear structure. X-13ARIMA-SEATS provides a wide range of diagnostic tools to assess these assumptions, and the outcomes are summarized in Figure 6.3.5.

Figure 6.3.5: Residuals

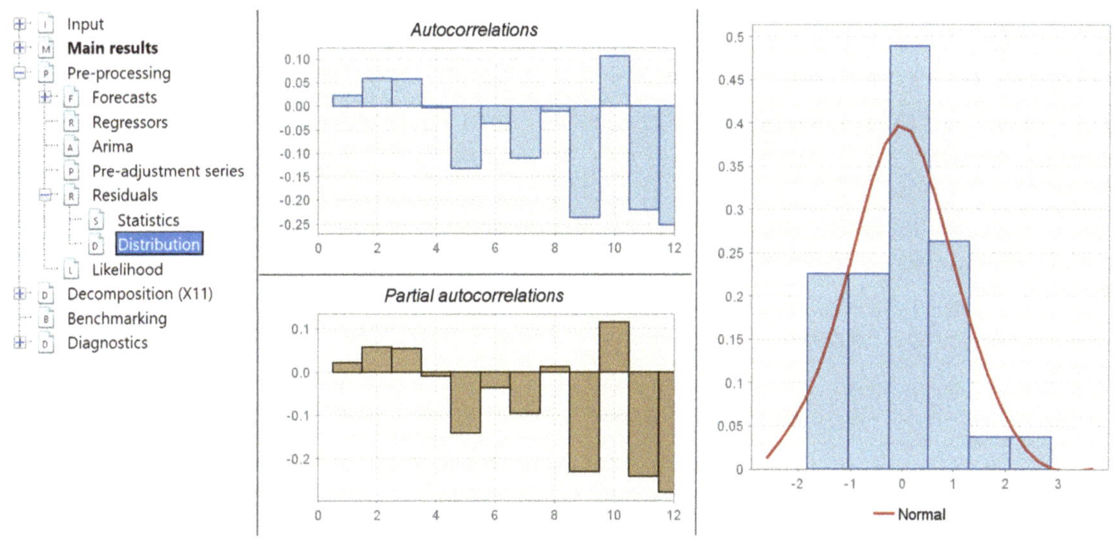

Source: Screenshot by the Asian Development Bank based on JDemetra+ 2.2.2.

Four main groups of tests are presented. First, normality is assessed using the Doornik-Hansen test,[18] alongside checks of skewness, kurtosis, and the mean of the residuals. Since prediction intervals are built on the assumption of normal errors, significant departures from normality reduce their reliability. Second, independence is evaluated using the Ljung-Box and Box-Pierce Q-statistics at both regular and seasonal lags, together with the Durbin-Watson statistic.[19] These tests examine whether autocorrelation remains in the residuals, which would indicate that systematic dynamics are left unexplained by the model. Third, randomness is examined with the Wald-Wolfowitz runs test, which checks whether positive and negative residuals occur in a sequence consistent with random variation.[20] Lastly, linearity is tested by applying the Ljung-Box and Box-Pierce statistics to squared residuals. Significant results here suggest time-varying variances or conditional heteroskedasticity, reducing the stability of the error process.[21]

18 J. A. Doornik and H. Hansen. 2008. An Omnibus Test for Univariate and Multivariate Normality. *Oxford Bulletin of Economics and Statistics*. 70 (s1). pp. 927–939.

19 G. E. P. Box and D. A. Pierce. 1970. Distribution of Residual Autocorrelations in Autoregressive-Integrated Moving Average Time Series Models. *Journal of the American Statistical Association*. 65 (332). pp. 1509–1526; G. M. Ljung and G. E. P. Box. 1980. Analysis of Variance with Autocorrelated Observations. *Scandinavian Journal of Statistics*. 7 (3). pp. 172–180; J. Durbin and G. S. Watson. 1950. Testing for Serial Correlation in Least Squares Regression: I. *Biometrika*. 37 (3–4). pp. 409–428.

20 A. Wald and J. Wolfowitz. 1940. On a Test Whether Two Samples Are from the Same Population. *Annals of Mathematical Statistics*. 11 (2). pp. 147–162.

21 R. F. Engle. 1982. Autoregressive Conditional Heteroscedasticity with Estimates of the Variance of United Kingdom Inflation. *Econometrica*. 50 (4). pp. 987–1007.

In addition to statistical tests, graphical diagnostics are provided to give a clearer picture. The sample autocorrelation function and partial autocorrelation function of the residuals are displayed to check whether any significant lags remain. In a well-specified model, these functions should show no systematic structure, since residuals are expected to resemble white noise. If lags appear, they point to unmodeled autocorrelation or moving-average components. Alongside these plots, a histogram of the residuals is shown with an overlaid normal density curve, which provides a visual check of the normality assumption. Ideally, the histogram bars should align closely with the bell-shaped curve, confirming that the residuals are approximately normal.

6.4 Decomposition

Once the pre-adjustment steps are complete, X-13ARIMA-SEATS proceeds with the decomposition of the time series into its main components: trend-cycle, seasonal, and irregular. This step lies at the core of seasonal adjustment, as it separates the long-term evolution of the series from short-term fluctuations and recurring seasonal patterns. The trend-cycle component captures the underlying direction of the series, reflecting long-term growth or decline as well as medium-term cyclical movements. The seasonal component represents regular patterns that repeat over fixed intervals, such as monthly or quarterly effects, while the irregular component reflects unsystematic noise or short-lived events that cannot be attributed to either trend or seasonality.

The results are organized into several tables that summarize different aspects of the decomposition output.

- **Part A** covers the prior adjustments, including corrections for significant known extremes and trading-day effects. These steps ensure that anomalies in the raw data do not distort the decomposition.
- **Part B** presents the first automatic correction of the series, where the irregular component is initially estimated, extreme values are detected and corrected, and trading-day adjustments are applied.
- **Part C** reports the second automatic correction, which further refines the estimation of the irregular component and rechecks for extreme observations and trading-day influences, enhancing the robustness of the results.
- **Part D** is devoted to the seasonal adjustment itself. Temporary seasonally adjusted series are first calculated (Tables D1–D6), followed by smoothing with the Henderson moving average and re-estimation of seasonal factors (Tables D7–D10). The final stage produces the definitive seasonally adjusted series (Table D11), along with the extracted trend-cycle (Table D12) and irregular component (Table D13). These outputs form the core of the final results most often used for analysis and policy purposes.
- **Part E** summarizes the components modified for large extreme values, documenting how exceptional observations were handled.

6.5 Quality Diagnostics

X-13ARIMA-SEATS provides a comprehensive set of diagnostics that enable analysts to judge whether a given specification has produced a reliable seasonal adjustment and whether the adjusted series can be confidently used for interpretation and publication. These diagnostics include both numerical results and qualitative messages that summarize the outcome of each test. Results are classified on a scale from **Good** to **Error** (Table 6.5.1). A **Good** rating indicates that the test has been successfully passed, while **Uncertain** suggests that further fine-tuning could improve the adjustment. Ratings such as **Bad**, **Severe**, or **Error** point to significant shortcomings and signal that the adjustment should be reconsidered.

Table 6.5.1: Messages of JDemetra+ for Quality Measures

Test Statistics	Reason
Undefined	The test is probably not designed or not suitable for that process, or an unsatisfied condition to calculate the statistics.
Error	The test statistic is unacceptable and needs to be recalculated.
Severe	The test statistic is unacceptable and needs to be recalculated.
Bad	The test statistic is critical and needs to be recalculated.
Uncertain	The test statistic is acceptable, but fine-tuning could remove the uncertainty.
Good	The test statistic is acceptable; no further attempt is needed.

Source: JDemetra+ Reference Manual.

The **Diagnostics** summary panel (Figure 6.5.1) brings these results together in a single view. It lists the outcomes of the basic checks, residual tests, M-statistics, and seasonality diagnostics, each highlighted in green, yellow, or red according to the message level. For instance, in the example shown, most indicators are **good**, confirming that the model specification and seasonal adjustment are reliable.

Figure 6.5.1: General View of Diagnostics

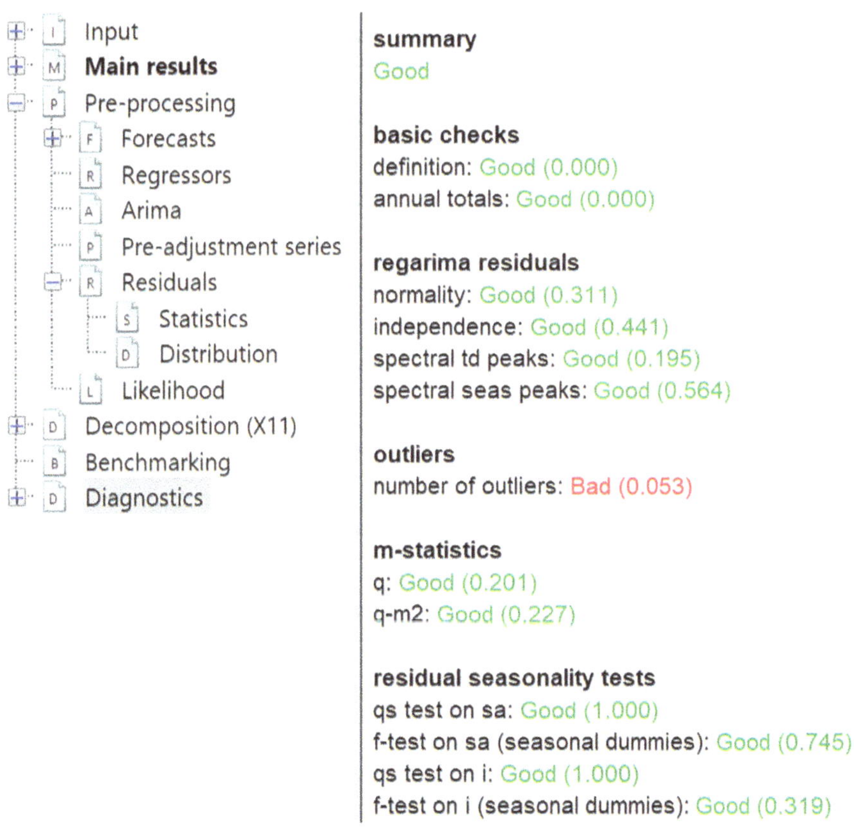

Source: Screenshot by the Asian Development Bank based on JDemetra+ 2.2.2.

However, the number of outliers is flagged as **bad**, suggesting that exceptional observations may need to be reviewed or handled differently. This summary allows users to immediately identify whether the adjusted series can be considered adequate or whether specific issues, such as residual seasonality, trading-day effects, or outliers, require further attention. More detailed analysis of each diagnostic is available in the subsequent subsections, where the individual tests are discussed in depth.

Spectral Analysis

The **Spectral Analysis** diagnostic evaluates whether seasonal or trading-day effects remain in the residuals, irregular components, or seasonally adjusted series (Figure 6.5.2). It displays the frequency spectrum of these series and highlights peaks at key points.[22] Blue lines mark seasonal frequencies; visible peaks here indicate that recurring seasonal patterns have not been fully removed, suggesting that the model specification, chosen filters, or estimation span may need adjustment. Red lines mark trading-day frequencies; peaks at these points signal inadequate treatment of calendar effects, such as trading-day regressors or leap-year adjustments, and often require revisiting how these factors are specified in the final model.

Figure 6.5.2: Spectral Analysis of Residuals

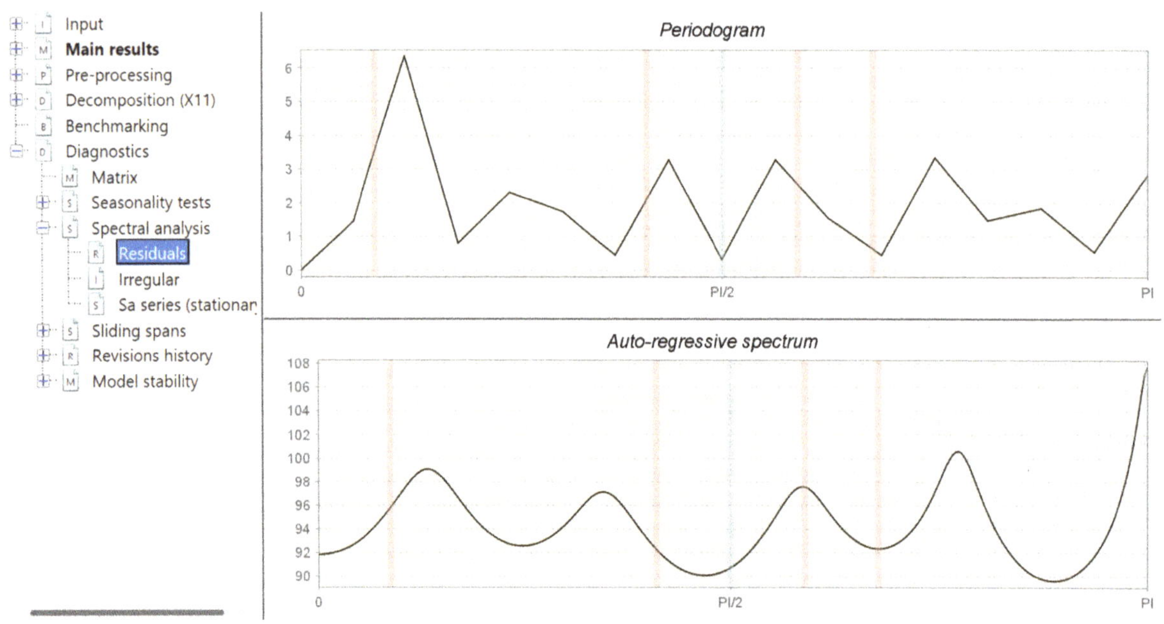

Source: Screenshot by the Asian Development Bank based on JDemetra+ 2.2.2.

[22] P. Bloomfield. 2004. *Fourier Analysis of Time Series: An Introduction.* 2nd ed. New York: John Wiley & Sons; M. B. Priestley. 1981. *Spectral Analysis and Time Series.* London: Academic Press; N. Beamish and M. B. Priestley. 1981. A Study of Autoregressive and Window Spectral Estimation. *Journal of the Royal Statistical Society: Series C (Applied Statistics).* 30 (1): 41–58.

Seasonality Tests

The **Seasonality Tests** diagnostics assess whether recurring seasonal patterns remain after adjustment. They are applied not only to the original series but also to the linearized series (with deterministic effects removed) and the residuals, so that any remaining seasonal structure can be detected at different stages of the modeling process.

Several complementary tests are used: autocorrelations at seasonal lags identify whether values are systematically linked across seasonal intervals; the Friedman and Kruskal–Wallis non-parametric tests check for distributional differences across seasons;[23] the periodogram test looks for strong peaks in the spectral density at seasonal frequencies; and regression-based tests with seasonal dummies verify whether fixed seasonal effects are still statistically significant.

Results are summarized in a diagnostic table (Figure 6.5.3), with clear indications of whether seasonality is present. In the example provided, all tests report that no residual seasonality remains in residuals, supporting the conclusion that the model and adjustment are adequate. If any of the tests had indicated the presence of residual seasonality, it would suggest that the adjustment was incomplete. In such cases, the analyst would need to revisit key modeling decisions: reconsidering the ARIMA specification, refining or expanding the set of regressors (for example, trading-day or holiday effects), or adjusting the time span used in estimation.

Figure 6.5.3: Seasonality Tests

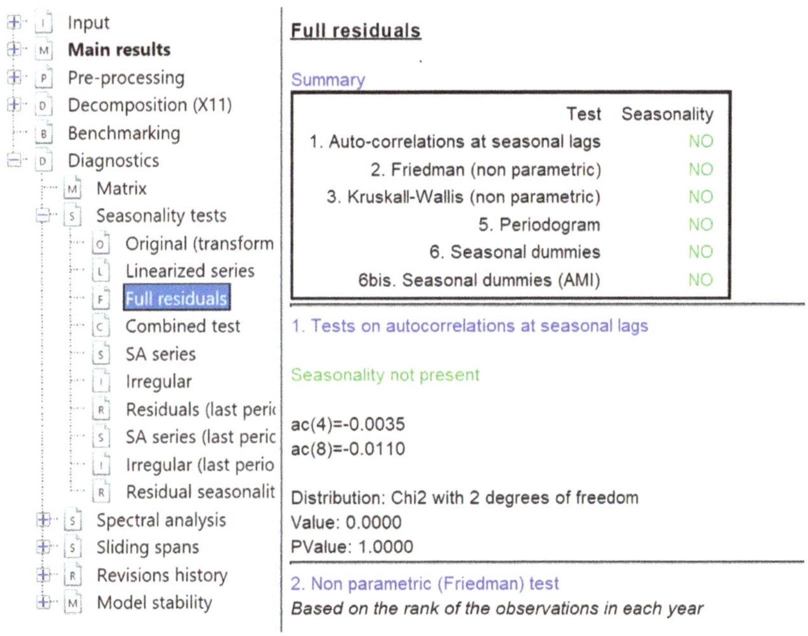

Source: Screenshot by the Asian Development Bank based on JDemetra+ 2.2.2.

[23] M. Friedman. 1937. The Use of Ranks to Avoid the Assumption of Normality Implicit in the Analysis of Variance. *Journal of the American Statistical Association.* 32 (200). pp. 675–701; W. H. Kruskal and W. A. Wallis. 1952. Use of Ranks in One-Criterion Variance Analysis. *Journal of the American Statistical Association.* 47 (260). pp. 583–621.

Sliding Spans

The **Sliding Spans** diagnostic, developed in X-13, is designed to test the stability of seasonal adjustment results over time. It is especially useful for series where seasonality may evolve or where outliers are frequent. The method divides the series into overlapping sub-periods, or *spans*, usually of 8 years. Each span is seasonally adjusted separately, and the results are compared to assess whether the estimated seasonal factors and adjusted series remain consistent across different samples.[24]

In the example shown in Figure 6.5.4, two spans are analyzed: 2013Q1–2021Q2 and 2014Q1–2022Q2. For each, the program computes the seasonal component, trading-day effect (if applicable), and seasonally adjusted series. Comparisons across spans highlight whether seasonal patterns are stable or subject to revision. The diagnostic uses several measures, including stable seasonality, moving seasonality, identifiable seasonality, and the Kruskal–Wallis test. By convention, differences larger than 3% are flagged as unstable, indicating that the seasonal component may not be robust. Abnormal values can be analyzed in detail on the relevant parts of the components.

Figure 6.5.4: Sliding Spans Analysis

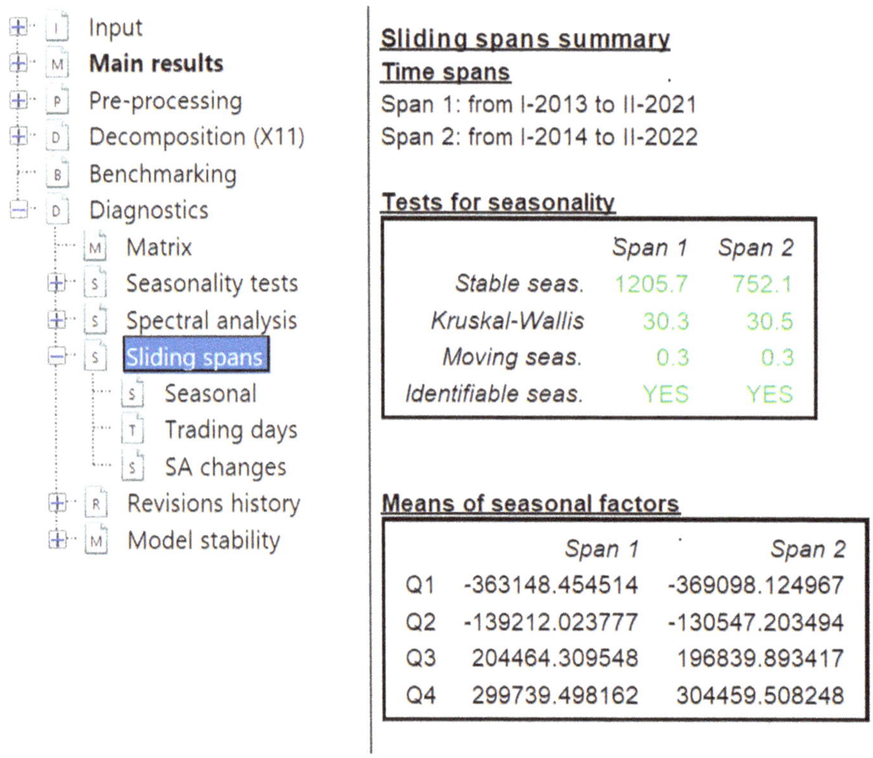

Source: Screenshot by the Asian Development Bank based on JDemetra+ 2.2.2.

[24] D. F. Findley, B. C. Monsell, and H. B. Shulman. 1990. Sliding Spans Diagnostics for Seasonal and Trend Stability. *Proceedings of the Business and Economic Statistics Section of the American Statistical Association*, 110–115.

Revision History

The **Revision History** diagnostic provides a systematic way to assess the stability of seasonal adjustment results when new data are added. Because official statistics are frequently updated, it is important to know whether these additions significantly alter previously published estimates. The tool addresses this by producing charts for the seasonally adjusted series, the trend-cycle component, and their growth rates, offering a clear view of how revisions affect different parts of the decomposition and whether the adjustment method remains robust as the dataset extends.

As illustrated in Figure 6.5.5, the revision history chart shows the evolution of adjustments across different time spans. Each blue circle represents the estimate obtained when the most recent observation at that time was treated as the end of the sample. The red line shows the current estimate using the full dataset. The process begins with a model estimated for the complete span, after which the span is gradually shortened and parameters are re-estimated. By default, only the model parameters are re-estimated, though the program can also perform a full re-estimation with outlier detection if required.

Figure 6.5.5: Revision History

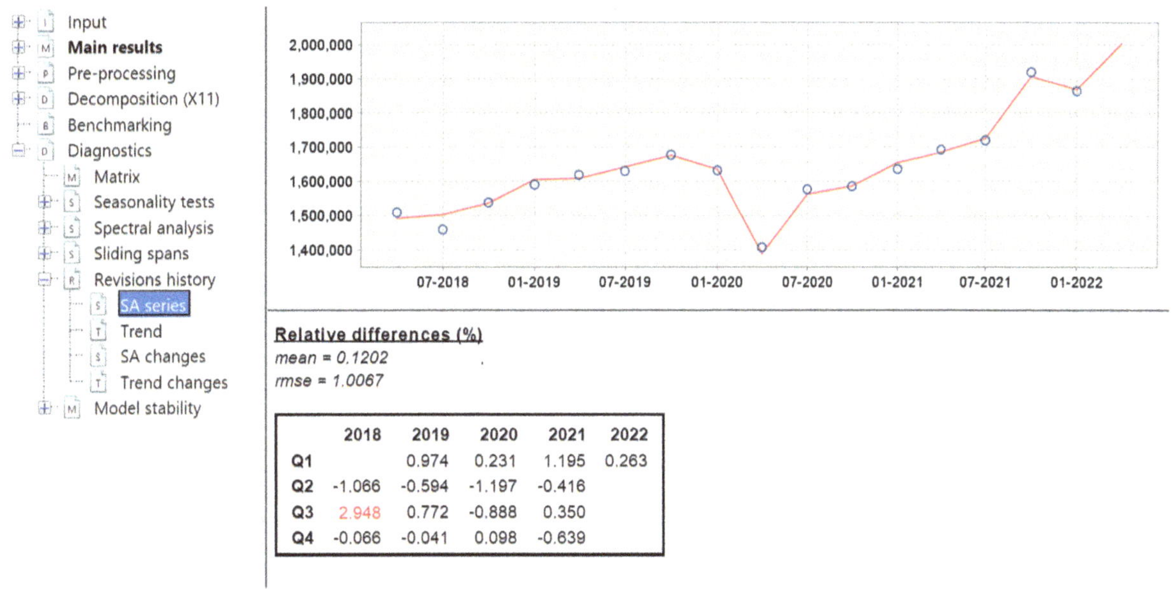

	2018	2019	2020	2021	2022
Q1		0.974	0.231	1.195	0.263
Q2	-1.066	-0.594	-1.197	-0.416	
Q3	2.948	0.772	-0.888	0.350	
Q4	-0.066	-0.041	0.098	-0.639	

Source: Screenshot by the Asian Development Bank based on JDemetra+ 2.2.2.

This analysis helps identify how sensitive the seasonal adjustment is to incoming data. If estimates for past periods change only slightly when new observations are added, the results can be considered stable and reliable. However, if revisions are large, it may suggest problems with the model specification, the presence of structural breaks, or instability in seasonal patterns. To quantify these effects, JDemetra+ also provides revision tables summarizing the differences between earlier and later estimates over the most recent 4 years. Depending on the decomposition type, these differences are presented either in absolute terms or as relative differences.

Model Stability

The final part of the **Diagnostics** module is the **Model Stability** section, which evaluates whether the estimated ARIMA parameters and regressor coefficients remain consistent when the model is applied to different sub-periods of the series. Stability is an essential feature of a reliable specification: if the model produces very different parameter values across time, the seasonal adjustment may not be robust, and future estimates could become unreliable.

The analysis is performed using a moving window of 8 years, which slides forward 1 year at a time. For each window, the ARIMA model is re-estimated, and the successive parameter values are recorded. These results are displayed in graphical form (Figure 6.5.6), with each point representing the estimate of a parameter at a given span. Stable results where parameters remain close to one another across windows indicate that the model is capturing the essential features of the data consistently. Shifts, however, may signal issues such as structural breaks in the series, an inadequate ARIMA specification, or omitted regressors.

Figure 6.5.6: Model Stability

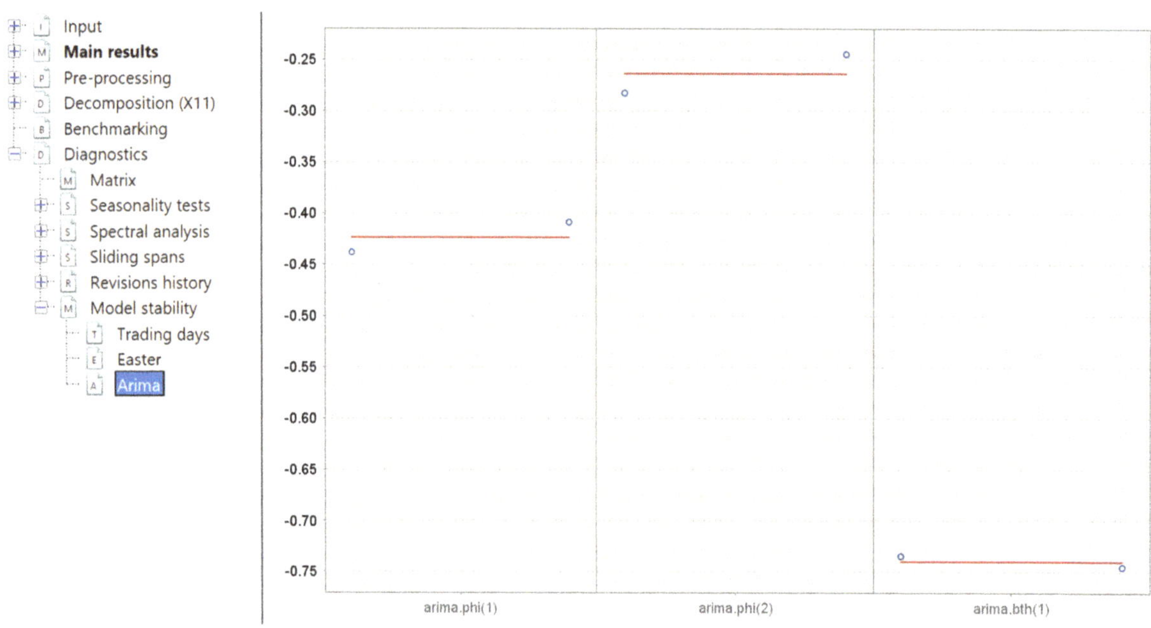

Source: Screenshot by the Asian Development Bank based on JDemetra+ 2.2.2.

M-Diagnostics

In addition to model-based diagnostics, JDemetra+ provides a set of M-statistics that summarize the quality of seasonal adjustment in a standardized way. These statistics, originally developed by the US Census Bureau for X-11 and extended in X-13, are widely used by statistical offices as benchmarks for assessing adjustment performance. They capture how well the decomposition separates the series into trend-cycle, seasonal, and irregular components, and whether the adjusted series behaves in a stable and interpretable manner.

As illustrated in Figure 6.5.7, the program reports 11 key M-statistics (M-1 through M-11), each measuring a specific property of the decomposition. Some statistics focus on the relative size of the irregular component (e.g., M-1 and M-2), while others examine the amount of autocorrelation or volatility in the irregular (M-3 to M-6). Additional measures evaluate the stability of seasonal factors over time (M-7 to M-9), as well as their behavior in recent years (M-10 and M-11). Together, these values provide a comprehensive picture of whether the adjustment has produced results that are both statistically sound and suitable for economic interpretation.

Figure 6.5.7: M-Diagnostics

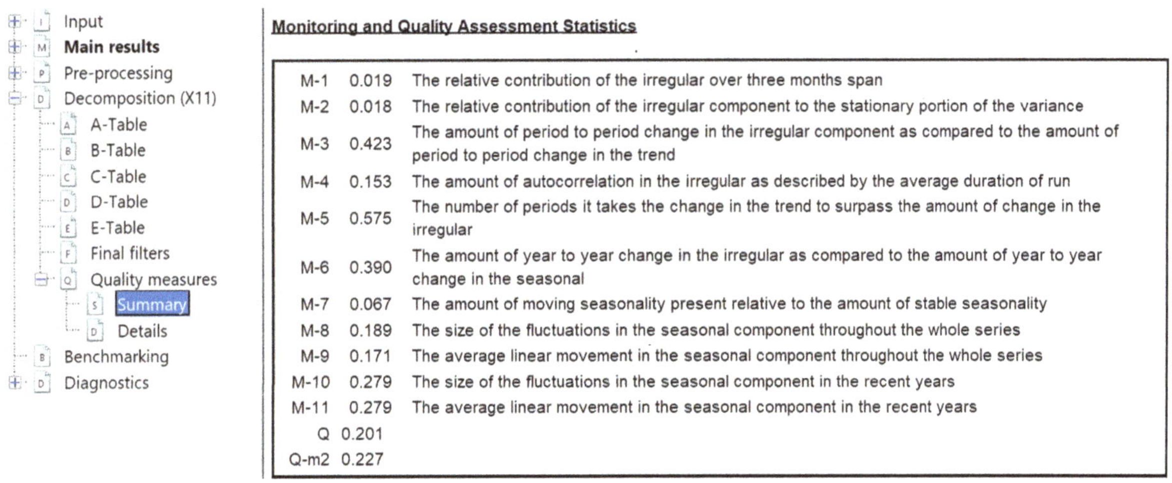

Source: Screenshot by the Asian Development Bank based on JDemetra+ 2.2.2.

To make interpretation easier, JDemetra+ also provides two summary indicators, Q and Q-M2, which condense the information from the full set of M-statistics into single measures of seasonal adjustment quality. Values near zero generally indicate a well-performing adjustment, while higher values point to possible issues such as residual seasonality or instability. In practice, analysts often compare these indicators across alternative model specifications, choosing the option that minimizes the statistics while still producing results that are economically meaningful. Beyond model selection, the M-statistics also serve as a tool for ongoing quality monitoring. A specification that performs well in the current sample may deteriorate as volatility, seasonal behavior, or irregular movements evolve over time. Regularly reviewing these measures helps ensure that adjustments remain robust and alerts users when a model needs to be refined or updated.

REFRESHING AND EXPORTING RESULTS 7

7.1 Refreshing Results

For official statistics, seasonal adjustment is not a one-time exercise but a recurring task. It is usually carried out on a monthly or quarterly basis whenever new data become available. While the underlying models are not re-estimated at every update, the adjustment itself should be re-run to incorporate the latest observations, which naturally leads to revisions in the seasonally adjusted series. In JDemetra+, updating an existing seasonal adjustment process with new data is straightforward, provided a few key conditions are met. The data source must first be imported using the **Providers** function, and the Workspace that contains the multi-processing setup should be stored in a dedicated folder whose path and name remain unchanged (Figure 7.1.1). It is also important to keep the name, location, and structure of the source file consistent from the beginning. Once these requirements are respected, the file can simply be updated with the new time series observations and any user-defined regressors, ensuring that the seasonal adjustment can be refreshed smoothly and without errors.

Figure 7.1.1: Saving Workspace

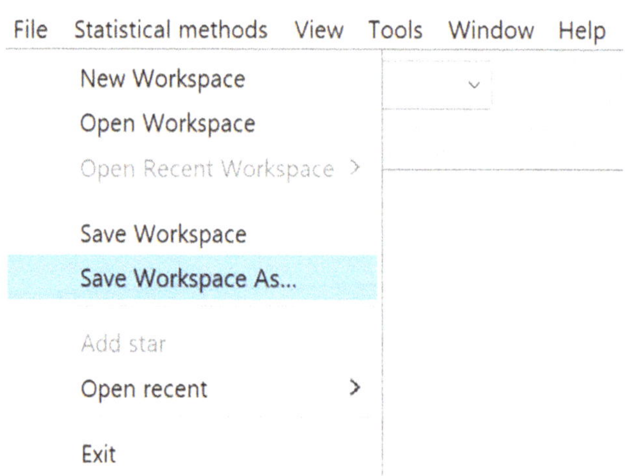

Source: Screenshot by the Asian Development Bank based on JDemetra+ 2.2.2.

The refreshment process can be applied either to a single series or to multiple series at once through a multi-processing setup. In both cases, the first step is to reload the previously saved Workspace (Figure 7.1.2, top panel), which restores all specifications, data links, and frozen results from earlier adjustments. For single processes, users can navigate to **Workspace/documents** and open the relevant document with a double-click, whereas for multi-processing, the corresponding file is accessed under **Workspace/multi-documents** in the same way.

Figure 7.1.2: Recall and Refresh Processing

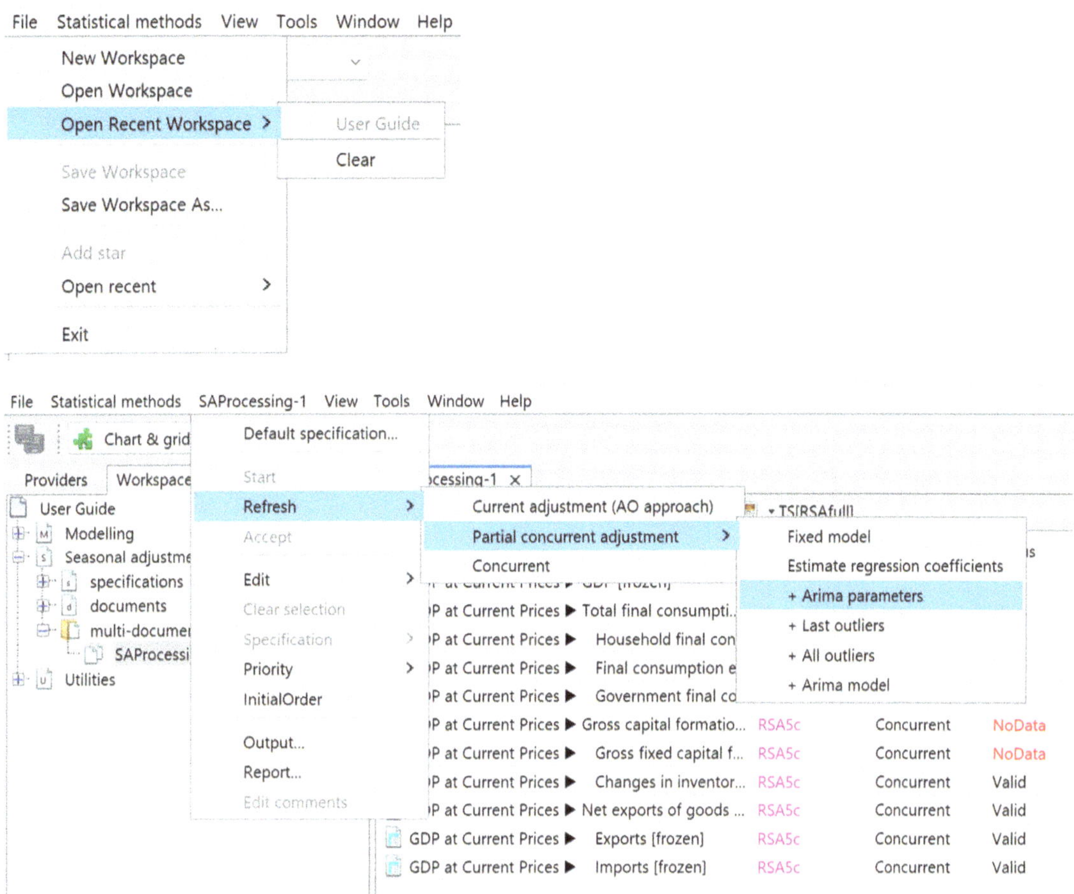

Source: Screenshot by the Asian Development Bank based on JDemetra+ 2.2.2.

Once the document is opened, the user should navigate to the processing tree in the **Workspace**. By selecting the relevant process (for example, SAProcessing-1), the seasonal adjustment can be refreshed through the **Refresh** option (Figure 7.1.2, bottom panel). At this stage, the user is prompted to choose between different refresh strategies, each offering a different balance between stability and flexibility.

- **Current adjustment (partial)** keeps the specification fixed and reuses pre-determined parameters and factors until the next scheduled review. This method minimizes revisions between review points but may leave residual seasonality if conditions change.
- **Partial concurrent adjustment** allows greater flexibility, as parameters, regressors, or outliers can be re-estimated whenever new or revised observations are added. Within this option, users can choose specific modes, such as re-estimating ARIMA parameters, identifying the most recent outliers, or re-checking all outliers.
- **Concurrent adjustment** provides the most comprehensive refresh, re-identifying the entire model each time new data are introduced. While this strategy produces more revisions, it ensures the seasonal adjustment fully reflects the most recent dynamics of the series.

7.2 Exporting Results and Metadata

For dissemination and further analysis, the results and metadata of seasonal adjustment should be exported in a clear and accessible format. In the case of a single analysis, the key summaries can be copied directly from JDemetra+. The overview pages under **Main results**, **Pre-processing**, **Decomposition**, and **Diagnostics** provide compact statistical information that can serve as reliable metadata, especially when accompanied by short interpretive notes on the quality of the adjusted series. For quick, ad-hoc exports of the actual time-series values, users can simply copy them from **Main results → Table**.

For more advanced workflows, especially when handling multi-processing, JDemetra+ offers built-in tools to export both results and metadata in a structured and reproducible way. The process begins from the document menu (for example, SAProcessing-1), where the user selects **Output** and specifies the preferred format (Figure 7.2.1). Within this menu, users can define the output path and file name, choose the layout of the tables, and decide which elements of the seasonal adjustment should be included. The software supports multiple formats such as .csv, .xls, and .txt, ensuring compatibility with a variety of statistical and analytical platforms. To provide full transparency, additional diagnostics and metadata, including test results and quality indicators, can be exported using the **CSV Matrix** option, which enables analysts to maintain a comprehensive set of evaluation tools for ongoing monitoring and validation.

Figure 7.2.1: Exporting Results and Metadata

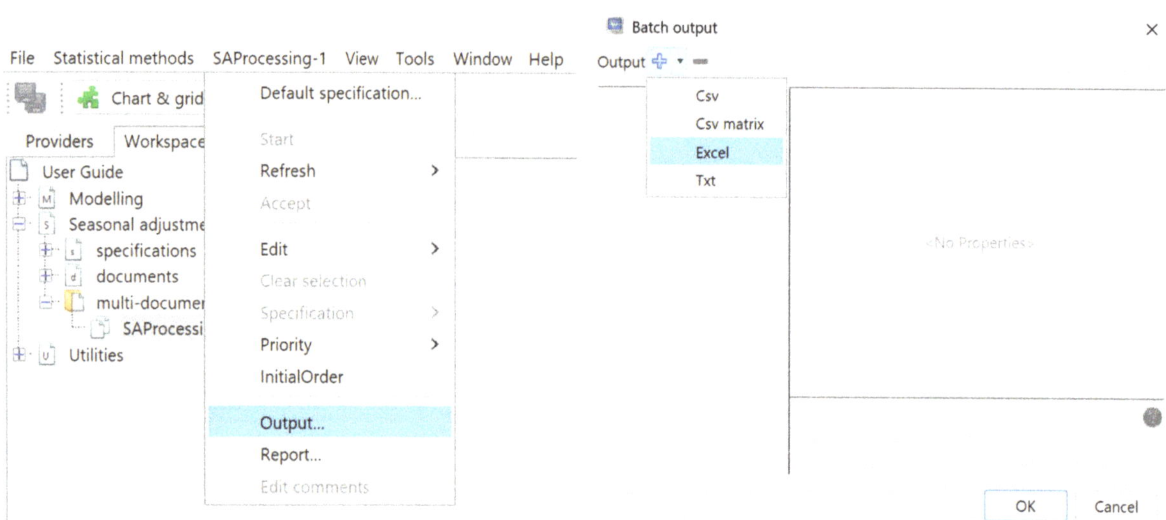

Source: Screenshot by the Asian Development Bank based on JDemetra+ 2.2.2.

USER COMMUNICATION 8

Seasonally adjusted data are among the most widely used outputs of official statistics, yet they are also among the most easily misunderstood. Users often focus on the numbers themselves without recognizing the technical choices, assumptions, and revisions that shape them. For this reason, communication must be treated as an integral part of the seasonal adjustment process. Best practice is to provide a clear framework that explains how the data are produced, why revisions occur, and what the results can and cannot tell us. In this way, user communication transforms seasonal adjustment from a purely technical exercise into a public service that supports trust, transparency, and effective decision-making.

Firstly, statistical offices should establish and communicate a clear and consistent seasonal adjustment policy. Such a policy should ensure that all series are treated with the same methods, that outliers and calendar effects are handled consistently, and that revisions follow a predictable schedule. Communicating this policy to users helps them understand the logic behind adjustments and anticipate changes over time. When users know, for instance, that models are re-estimated once a year, but adjustments are refreshed each quarter, they are better able to interpret revisions as part of a structured process rather than arbitrary changes.

Secondly, offices should provide documentation that is carefully tailored to different audiences. Internal documentation should be sufficiently detailed to guarantee reproducibility and continuity, including step-by-step procedures, model specifications, and explanations for the treatment of holidays and outliers. External documentation, by contrast, should be concise and accessible, offering clear explanations of the methods used, the revision policy, and key quality measures. Visual tools such as charts and tables should be used to improve readability and to prevent misinterpretation, for example confusing a revision with an actual economic change.

Lastly, statistical offices should ensure that continuous monitoring and quality control are part of the communication process. Diagnostics generated by JDemetra+, such as residual tests, spectral analysis, sliding spans, and revision history provide essential checks on model adequacy, but their meaning should be translated into plain language. Routine monitoring should be used to detect potential issues quickly, while annual reviews should assess whether changes in volatility or seasonal patterns require a revision of the models. The results of these diagnostics should be communicated in reports that are comprehensive for internal experts but simplified for external users, thereby reassuring all audiences that the seasonal adjustment process remains robust and reliable.

REFERENCES

N. Beamish and M. B. Priestley. 1981. A Study of Autoregressive and Window Spectral Estimation. *Journal of the Royal Statistical Society: Series C (Applied Statistics).* 30 (1). pp. 41–58.

W. R. Bell and S. C. Hillmer. 1984. Issues Involved with the Seasonal Adjustment of Economic Time Series. *Journal of Business & Economic Statistics.* 2 (4). pp. 291–320.

P. Bloomfield. 2004. *Fourier Analysis of Time Series: An Introduction.* 2nd ed. New York: John Wiley & Sons.

G. E. P. Box and D. R. Cox. 1964. An Analysis of Transformations. *Journal of the Royal Statistical Society: Series B (Methodological).* 26 (2). pp. 211–243.

G. E. P. Box and G. M. Jenkins. 1976. *Time Series Analysis: Forecasting and Control.* San Francisco: Holden-Day.

G. E. P. Box and D. A. Pierce. 1970. Distribution of Residual Autocorrelations in Autoregressive-Integrated Moving Average Time Series Models. *Journal of the American Statistical Association.* 65 (332). pp. 1509–1526.

C. Chen and G. C. Tiao. 1990. Random Level-Shift Time Series Models, ARIMA Approximations, and Level-Shift Detection. *Journal of Business & Economic Statistics.* 8 (1). pp. 83–97.

P. A. Cholette and E. B. Dagum. 1994. Benchmarking Time Series with Autocorrelated Survey Errors. *International Statistical Review.* 62 (3). pp. 365–377.

E. B. Dagum and S. Bianconcini. 2016. *Seasonal Adjustment Methods and Real-Time Trend-Cycle Estimation.* Cham: Springer.

E. B. Dagum and P. A. Cholette. 2006. *Benchmarking, Temporal Distribution, and Reconciliation Methods for Time Series.* New York: Springer.

J. A. Doornik and H. Hansen. 2008. An Omnibus Test for Univariate and Multivariate Normality. *Oxford Bulletin of Economics and Statistics.* 70 (s1). pp. 927–939.

J. Durbin and G. S. Watson. 1950. Testing for Serial Correlation in Least Squares Regression: I. *Biometrika.* 37 (3–4). pp. 409–428.

R. F. Engle. 1982. Autoregressive Conditional Heteroscedasticity with Estimates of the Variance of United Kingdom Inflation. *Econometrica.* 50 (4). pp. 987–1007.

Eurostat. 2018. *Handbook on Seasonal Adjustment.* 2018 Edition. Luxembourg: Publications Office of the European Union. https://ec.europa.eu/eurostat/documents/3859598/8939616/KS-GQ-18-001-EN-N.pdf.

———. 2024. *ESS Guidelines on Seasonal Adjustment – 2024 Edition.* Luxembourg: Publications Office of the European Union.

D. F. Findley, B.C. Monsell, and H. B. Shulman. 1990. Sliding Spans Diagnostics for Seasonal and Trend Stability. *Proceedings of the Business and Economic Statistics Section of the American Statistical Association,* 110–115.

D. F. Findley, B. C. Monsell, W. R. Bell, M. C. Otto, and B.-C. Chen. 1998. New Capabilities and Methods of the X-12-ARIMA Seasonal-Adjustment Program. *Journal of Business & Economic Statistics.* 16 (2). pp. 127–152.

D. F. Findley and B. C. Monsell. 2009. Model Selection and Outlier Detection in Seasonal Adjustment. *Journal of Official Statistics.* 25 (1). pp. 35–70.

M. Friedman. 1937. The Use of Ranks to Avoid the Assumption of Normality Implicit in the Analysis of Variance. *Journal of the American Statistical Association.* 32 (200). pp. 675–701.

V. Gómez and A. Maravall. 1995. *Programs TRAMO and SEATS.* Florence: European University Institute.

V. Gómez, A. Maravall, and D. Peña. 1999. Missing Observations in ARIMA Models: Skipping Strategy versus Additive Outlier Approach. *Journal of Econometrics.* 88 (2). pp. 341–363.

S. Grudkowska. 2015. *JDemetra+ User Guide.* Warsaw: National Bank of Poland.

S. Hillmer and G. C. Tiao. 1982. An ARIMA-Model-Based Approach to Seasonal Adjustment. *Journal of the American Statistical Association.* 77 (377). pp. 63–70.

International Monetary Fund. 2017. *Quarterly National Accounts Manual.* Washington, DC: IMF.

W. H. Kruskal and W. A. Wallis. 1952. Use of Ranks in One-Criterion Variance Analysis. *Journal of the American Statistical Association.* 47 (260). pp. 583–621.

D. Ladiray and B. Quenneville. 2001. *Seasonal Adjustment with the X-11 Method.* New York: Springer.

G. M. Ljung and G. E. P. Box. 1980. Analysis of Variance with Autocorrelated Observations. *Scandinavian Journal of Statistics.* 7 (3). pp. 172–180.

A. Maravall and D. Pierce. 1987. A Prototypical Seasonal Adjustment Model. *Journal of Time Series Analysis.* 8 (2). pp. 177–193.

B. C. Monsell. 2009. Update on the Development of X-13ARIMA-SEATS. *Proceedings of the Joint Statistical Meetings: American Statistical Association.* 13: 1–10.

D. Peña, G. C. Tiao, and R. S. Tsay. 2001. *A Course in Time Series Analysis.* New York: John Wiley & Sons.

M. B. Priestley. 1981. *Spectral Analysis and Time Series.* London: Academic Press.

B. Quenneville, P. A. Cholette, E. B. Dagum, and B. Rioux. 2004. Adjustment of Seasonally Adjusted Series to Annual Totals. *Proceedings of the American Statistical Association, Business and Economic Statistics Section,* 1–8.

United States Census Bureau. 2023. *X-13ARIMA-SEATS Reference Manual.* Washington, DC: US Department of Commerce.

A. Wald and J. Wolfowitz. 1940. On a Test Whether Two Samples Are from the Same Population. *Annals of Mathematical Statistics.* 11 (2). pp. 147–162.

www.ingramcontent.com/pod-product-compliance
Lightning Source LLC
Chambersburg PA
CBHW041759220926
51847CB00014B/254
* 9 7 8 9 2 9 2 7 7 6 1 1 4 *